Global Warming

To George Lockwood
and the spirit of curiosity

Global Warming
Understanding the forecast

David Archer
University of Chicago

BLACKWELL PUBLISHING
350 Main Street, Malden, MA 02148-5020, USA
9600 Garsington Road, Oxford OX4 2DQ, UK
550 Swanston Street, Carlton, Victoria 3053, Australia

3 2008

Library of Congress Cataloging-in-Publication Data

Archer, David, 1960-
Global warming: understanding the forecast/David Archer.
p. cm.
Includes bibliographical references and index.
ISBN: 978-1-4051-4039-3 (pbk. : acid-free paper)
1. Global warming. 2. Global temperature changes. 3. Greenhouse effect, Atmospheric.
4. Global warming–Political aspects. 5. Global warming–Economic aspects.
I. Title

QC981.8.G56A73 2007
551.6–dc22

2006009415

A catalogue record for this title is available from the British Library.

Set in 10.5/12.4 Minion
by Newgen Imaging Systems (P) Ltd., Chennai, India
Printed and bound in Singapore by C.O.S. Printers Pte Ltd

The publisher's policy is to use permanent paper from mills that operate a sustainable forestry policy, and which has been manufactured from pulp processed using acid-free and elementary chlorine-free practices. Furthermore, the publisher ensures that the text paper and cover board used have met acceptable environmental accreditation standards.

For further information on
Blackwell Publishing, visit our website:
www.blackwellpublishing.com

Contents

Online models

A model of infrared radiation in the atmosphere
http://understandingtheforecast.org/Projects/infrared_spectrum.html

A model of visible + infrared radiation in the atmosphere http://understandingtheforecast.org/Projects/full_spectrum.html

A model of the geological carbon cycle
http://understandingtheforecast.org/Projects/geocarb.html

ISAM Integrated assessment model for future climate change
http://understandingtheforecast.org/Projects/isam.html

A Hubbert's Peak calculator
http://understandingtheforecast.org/Projects/hubbert.html

The Kaya Identity model for the growth of the human footprint
http://understandingtheforecast.org/Projects/kaya.html

Browsing the results of a coupled climate model
http://understandingtheforecast.org/Projects/bala.html

Preface

Global Warming: Understanding the forecast is based on class lectures for undergraduate nonscience majors at the University of Chicago, developed by Ray Pierrehumbert and myself. The class serves as partial fulfillment of our general education or "core" science requirements. We teach the class, and I have written this textbook, in a mechanistic way. We aim to achieve an intuitive understanding of the ropes and pulleys of the natural world, a fundamental scientific foundation that will serve the student longer than would a straight presentation of the latest predictions.

The text aims at a single problem – assessing the risk of anthropogenic climate change. The story ranges from science to economics to policy, through physics, chemistry, biology, geology, and of course atmospheric science. We see the distant past and the distant future. In my opinion, by looking at a problem from many angles, the student gets a pretty decent view of how a working scientist really thinks. This is as opposed to, say, taking a survey tour of some scientific discipline.

The text is suitable for students of all backgrounds. We do make some use of algebra, mostly in the form of what are known as (gasp) story problems. The student will be exposed to bits and pieces of chemistry, physics, biology, geology, atmospheric science, and economics, but no prior knowledge of any of these topics is required. One can learn something of what each field is about by learning what it is good for, within the context of the common unifying problem of global warming.

I have provided a project associated with each chapter after the first, either a computer lab or a paper-and-pencil exercise, suitable to do in lab sessions or as homework. The first three are paper-and-pencil exercises, aimed at building a foundation for understanding the computer lab exercises that follow. The models run on our computers at the University of Chicago, which can be accessed through web pages. No special setup of the computer is required and students can work just as equally well in a computer lab as in Starbucks (actually, it would be interesting to see if that's true).

This book has benefited through thoughtful reviews by Andy Ridgwell, Stefan Rahmstorf, Gavin Schmidt, and a fourth anonymous reviewer. The web interface to the models benefited from inputs by Jeremy Archer. The visible/infrared radiation model was constructed by Ray Pierrehumbert and Rodrigo Caballero. The ISAM carbon cycle model was provided by Atul Jain. The back cover photo of the glacier was taken by Lonnie Thompson. The project in Chapters 11 and 12 makes use of model output provided by G. Bala, and was plotted using ferret, developed at the NOAA PMEL.

Instructors may request a CD with solutions and artwork from the book via this email address: artworkcd@bos.blackwellpublishing.com

1
Humankind and climate

Everyone always complains about the weather, but no one ever does anything about it.

Mark Twain

The glaciers are melting, but is it us?

Is it really possible that human activity could alter the weather? As I write, it is a crisp, clear fall day. What would be different about this day in a 100 years, in a world where the chemistry of the atmosphere has been altered by human industrial activity?

There is no doubt that the Earth is warming. Mountain glaciers are disappearing. The Arctic coast is melting. The global average temperature records are broken year after year. The growing season has been getting longer. Plans are being made to abandon whole tropical islands as they sink into the Pacific Ocean. Shippers are awaiting the opening of the Northwest Passage that early explorers searched for in vain, with the melting of the sea ice in the Arctic. The Atlantic has had so many hurricanes this season that we are running out of letters in the English alphabet for names, having to resort to Greek letters – we are now up to epsilon.

Of course, the world has variable weather all by itself, naturally. Is it likely that some of our recent weather has been impacted by human-induced climate change, or how much of this would have happened anyway? If humans are indeed changing the climate, do we know if this is a bad thing? How does the future evolution of climate compare with the climate impacts we may be seeing today?

Weather versus climate

We should distinguish at the outset between climate and weather. Weather is chaotic, which means that it cannot be forecast very far into the future. Small errors in the forecast grow with time, until eventually the forecast is nothing but an error. By climate, we mean some sort of an average of the weather, say averaged over 10 years, more or less. We cannot predict the details of rain versus shine on Tuesdays versus Saturdays very far into the future, but we can hope to forecast the average rainfall in some location at some time of year. Weather is chaotic, but by taking the average, we arrive at something that is not chaotic, which seems to be in some ways predictable. We will return to this topic in Chapter 6.

Human-induced changes in climate are expected to be small when compared with the variability associated with weather. Temperature in the coming century is projected to rise by a few degrees centigrade (Chapter 12). This is pretty small compared with the temperature differences between the equator and the poles, between winter and summer, or even between day and night. One issue this raises is that it is trickier to discern a change in the average when the variability is so much greater than the trend. Careers are spent computing the global average temperature trend from the 100+ year thermometer record (Chapter 11).

A small change in the average relative to a huge variability also raises the question of whether a change in the average will even be noticeable. One way that the average weather matters is in precipitation. Ground water tends to accumulate, reflecting rainfall over the past weeks and months and years. It may not matter to a farmer whether it rains on a Tuesday or Saturday, but if the average rainfall in a region changes that could spell the difference between productive and nonproductive farming. A change in the average climate will change the growing season, the frequency of extreme hot events, the distribution of snow and ice, the optimum growth localities of plants and agriculture, and the intensity of storms.

In addition to day-to-day weather, there are longer-lasting variations in climate. One past climate regime was the Little Ice Age, approximately 1650–1800, bringing variable weather, by some records 1°C colder on average, to Europe. Before that was the Medieval Optimum, perhaps 0.5°C warmer over Europe, coincident with a prolonged drought in the American southwest. We will discuss the causes of these climate changes in Chapter 11, but for now it is enough to observe that relatively small-sounding average-temperature shifts produced noticeable changes in human welfare and the evolution of history. The climate of the Last Glacial Maximum, 20,000 years ago, was so different from today that the difference would be obvious even from space, and yet the average temperature difference between then and today is only about 5–6°C (Chapter 8). Another implication of these natural climate shifts is that it makes it more difficult to figure out whether the present-day warming is natural or caused by rising greenhouse gas concentrations and other human impacts on climate.

Forecasting climate change

The fundamental process that determines the temperature of the Earth is the balance between energy flowing to the Earth from the Sun and energy flowing away from the Earth into space. Heat loss from Earth depends on the Earth's temperature, among other things (Chapter 2). A hotter Earth loses heat faster than a cooler one, all else being equal. The earth balances its energy budget by warming up or cooling down, finding the temperature at which the energy fluxes balance, with outflow equaling inflow. The feedback is analogous to a sink with water flowing in from a faucet. The faucet fills the sink at some constant rate, while outflow down the drain depends on the water level in the sink. The sink fills up until water drains out as fast as it comes in (Chapter 3).

The increase in outgoing energy with increasing temperature of the Earth results in a feedback which stabilizes the temperature of the Earth so that the energy fluxes in and out balance each other (Chapter 7). It is possible to change the average temperature of the Earth by altering the energy flow either coming in or going out. In our sink, one way to raise the water level is to turn up the faucet and wait a few minutes. The water will rise until it finds a new equilibrium water depth. We can also alter the water level by partly constricting the drain. Egg shells and orange peels work well for this purpose. If the drain is partially obstructed, the equilibrium water level will rise.

The incoming energy to Earth might change if the Sun changes its brightness. It is known that there is a small variation in the brightness of the Sun correlated with the number of sunspots. Sometimes sunspots disappear altogether, presumably changing the solar output. The Maunder Minimum was such a period, 1650–1700, and coincided with the Little Ice Age.

Some of the incoming sunlight is reflected back to space without ever being absorbed (Chapter 7). It we somehow make the Earth brighter, more reflective, it will tend to cool. Clouds reflect light, and so does snow. Bare soil in the desert reflects more light than vegetation does. Smoke emitted from coal-burning power plants contains particles that can reflect light. Thus, changes in cloudiness, land cover, and smoke can change climate.

The climate-forcing agent at the heart of the global warming problem is the greenhouse effect from rising CO_2 concentration in the atmosphere. CO_2, a greenhouse gas, makes it more difficult for energy leaving the Earth to escape to space. Water vapor and methane are also greenhouse gases. Most of the gases in the atmosphere are not greenhouse gases, but are completely transparent to infrared (IR) light. The outgoing energy from the Earth passes through them as if they were not there. Greenhouse gases have the ability of absorb and emit IR light. The impact they have on climate depends on their concentration because the more the gas, the more the light absorbed (Chapter 4). The strength of the greenhouse effect also depends on the temperature structure of the atmosphere (Chapter 5).

Water vapor is a tricky greenhouse gas because the amount of water vapor in the atmosphere is determined by climate. Water tends to evaporate when the air is warm and condense as rain or snow when the air is cool. Water vapor, it turns out, amplifies the warming effects from changes in other greenhouse gases. This water vapor feedback doubles or even triples the temperature change we expect from rising atmospheric CO_2 concentration. Clouds are very effective at absorbing and emitting IR light, acting like a completely impenetrable greenhouse gas. A change in cloudiness also affects the incoming visible light energy flux by reflecting it (Chapter 7).

Human activity has the potential to alter climate in several ways. Rising CO_2 concentration from combustion of fossil fuel is the largest and longest lasting human-caused climate forcing agent, but there are other greenhouse gases, such as methane (CH_4) and other hydrocarbon molecules, nitrous oxide, and ozone, the concentrations of which are also changing because of human activities. Particles from smoke stacks and internal combustion engines reflect incoming visible light, altering the heat balance. Particles in otherwise remote clean air may change the average size of cloud droplets, which has a huge but very uncertain impact on sunlight reflection (Chapter 10).

Many of these climate drivers themselves respond to climate, leading to stabilizing or destabilizing feedbacks. There are several examples where the prehistoric climate record shows more variability in climate than models tend to predict, presumably because there exist positive feedbacks in the real world that are missing in the models. For example, climate cools, so a forest changes to a tundra, allowing more of the incoming sunlight to be reflected to space, cooling the climate even more. A naïve model with forests that does not respond to climate would underestimate the total amount of cooling. In the global warming forecast, the feedbacks are everything. The forecast would be much easier in a simpler world (Chapter 7).

The forecast for the coming century is also tricky because some parts of the climate system take a long time to change, such as the melting of an ice sheet or the warming of a deep ocean. Coming back to our sink analogy, not only do we have to estimate the eventual change in the water level in the sink, but we also need to predict how quickly it will rise. For the sink this is not so bad whereas for climate it makes prediction considerably harder (Chapter 12).

Carbon, energy, and climate

Climate change from fossil fuel combustion is arguably the most challenging environmental issue we face because CO_2 emission is at the heart of how we produce energy, which is pretty much at the heart of our modern standard of living. The agricultural revolution, which supports a human population of 6 billion people and hopefully more, has at its heart the industrial production of fertilizers, a very energy-intensive process. It's not easy to stop CO_2 emission, and countries and companies that emit lots of CO_2 have a strong interest in continuing to do so (Chapter 9).

The energy we extract from fossil fuels originated in the nuclear fires of the Sun. Visible light carried the energy to Earth, where it was converted by photosynthesis in plants into chemical energy in chemical bonds of carbon, hydrogen, oxygen, and other elements. Plants have two motives for doing this, one to store energy and the other to build CO_2 molecules from the atmosphere into the biochemical machinery of life (Chapter 8).

Most of the biological carbon we use for fossil fuels was photosynthesized millions of years ago. Over geologic time, some of the biological carbon has been converted into the familiar fossil fuels such as oil, natural gas, and coal. Coal is the most abundant of these, while the types of oil and gas that are currently being extracted will be depleted in a few decades (Chapter 9). Stored carbon energy is used to do work, in plants, animals, and now in automobiles, by reacting the carbon with oxygen to produce CO_2. In living things this process is called respiration, explaining why we need to breathe (to obtain oxygen and get rid of CO_2) and eat (to get biological carbon compounds) (Chapter 8).

CO_2 is released into the atmosphere to join the beautiful cacophony that is the carbon cycle of the biosphere. Trees and soils take up and release carbon, as does the ocean. Cutting of tropical forests releases CO_2 into the atmosphere, while forests in the high latitudes appear to be taking up atmospheric CO_2. Most of the CO_2 we release into the atmosphere will eventually dissolve in the ocean, but this process takes several centuries. A small fraction, about 10%, of the CO_2 released will continue to alter climate for hundreds of thousands of years into the future (Chapter 10).

Assessing the risk

Is mankind creating a global warming trend? We can try to answer this question by comparing the history of the Earth's temperature with the history of the different reasons why temperature might have changed, what we call climate forcings. The Sun is more intense at some times than others. Volcanos occasionally blow dust into the stratosphere where it reflects sunlight back to space. Greenhouse gases and smokestack aerosols are two anthropogenic climate forcings. The conclusions we will come to are, first, that it is getting warmer and, second, that it is easy to explain the warming as caused by increased greenhouse gas concentrations, but impossible to explain them as occurring naturally (Chapter 11).

The forecast for the climate of the coming century is for a temperature increase of 2–5°C by the year 2100. The amount of warming depends not only on the intensity of feedbacks with water vapor and clouds, but also on the time-dependent evolution of the climate system as it changes from one state to another. The feedbacks affect not only the intensity but also the response time of climate to the new greenhouse gas concentrations. An increase in temperature of a few degrees does not sound catastrophic, and for some parts of the world it may not be. We can get an idea of the effect of such a temperature change by looking into the past, at natural climate shifts that occurred through the last 1000 years, perhaps 0.5–1°C changes, or at the end of the glacial maximum, a warming of about 6°C on global average. Temperature changes as small as the 0.5°C medieval warm time in Europe were associated with prolonged drought in the American southwest, of sufficient intensity to spell the end of the Mayan civilization (Chapter 12).

When we ask questions about the impact of global warming on humanity, we begin to enter the realm of the social sciences, especially of economics. Using models, economists can forecast the future just as meteorologists or climatologists can. One approach is to compare the predicted costs of predicted climate change against the predicted costs of avoiding climate change. It is of course difficult to put a monetary value on the natural world, but for what it is worth the projections are that reducing CO_2 emissions substantially might cost a few percentage of the global net economic production. That certainly would be a lot of money if you were looking at it piled in a heap, but in an economy that is growing by a few percent per year, a cost of a few percent per year would only set the trajectory of economic production back by a year or two (Chapter 13).

Economics is an awkward tool for use in global warming decision making, however. For one thing, it seems crass to represent the demise of the natural world, biodiversity, and wilderness as a simple monetary cost. Economics is not well suited to comparing costs across long stretches of time because of a concept called the discount rate. Finally, economic forecasts in general are trickier than climate forecasts because the underlying laws of economics are not as simple and immutable as they are in the physical world. Ultimately, the question of climate change may be a matter of ethics and fairness, as much as one of profit and loss.

Greenhouse gas emission to the atmosphere is an example of a situation called the tragedy of the commons. The benefits of fossil fuel combustion go to the individual whereas the costs of climate change are paid by everyone. In this situation there is a natural tendency for everyone to overexploit the commons. The solution to this is some form of collective regulation. International negotiations have resulted in an agreement called the Kyoto Protocol, which aims to limit emissions of CO_2 gas. The Kyoto Protocol, if successful, would curtail emissions by about 6% below 1990 levels, while carbon cycle models show that eventual cuts of order 50% would be required to truly stabilize the CO_2 concentration of the atmosphere.

Carbon emissions could be lessened quite a bit by conservation and efficiency. Sources of carbon-free energy such as wind and solar energy are rapidly being scaled up. A forecast of the amount of energy that civilization will require in the coming century suggests that large, new sources of energy may be required in the future. Nuclear energy is essentially carbon-free but it would take a new nuclear power plant of current design built every other day for the next 100 years to keep up with the forecast energy demand. New ideas include solar cells on the Moon, beaming energy back to Earth as microwave radiation, or high-altitude windmills, mounted on kites tethered in the jet stream.

With our growing technological and intellectual prowess, as well as our exploding population, we are inevitably taking over the job of managing the biosphere. May we do it wisely!

Part I

The greenhouse effect

2
Blackbody radiation

Energy travels through a vacuum between the Sun and the Earth by means of electromagnetic radiation, or light. Objects can absorb energy from light, and they can also emit light energy, if the vibrations of their chemical bonds generate oscillations of the electrical field. An object that can emit all wavelengths of light is called a blackbody. The spectrum of light from a blackbody sparked the development of quantum mechanics, a revolution in physics and natural philosophy.

Heat and light

Heat is simply the bouncing-around energy of atoms. Atoms in gases and liquids fly around, faster if it is hot and slower if it is cold. Atoms with chemical bonds between them stretch, compress, and bend those bonds, again more energetically at a high temperature. Perhaps you knew this already, but have you ever wondered how we can feel such a thing? Atoms are tiny things. We can't really feel individual atoms. But it doesn't take any state-of-the-art laboratory technology to tell how fast the iron atoms in your stove burner are bouncing around. All you have to do is touch it, if you dare. Actually, another method may be occurring to you, that you could look at it; if it's hot, it will glow. It glows with blackbody radiation, which we'll get back to later.

You can feel the hot stove because the energetic bounciness of the stove atoms gets transferred to the atoms in your nerves in your fingers. The fast-moving atoms of the burner bounce off the atoms in your finger, and the fast ones slow down a bit and the slower ones bounce back with a little more energy than they started with. Biological systems have evolved to pay attention to this, which is why you can feel it, because too much energy in our atoms is a dangerous thing. Chemical bonds break when they are heated too much, and we could cook ourselves. Burning your finger by touching a hot electric stove burner is an example of heat ***conduction***, the easiest form of heat flow to imagine.

Energy through a vacuum

A thermos bottle is designed to slow the flow of heat through its walls. You can put warm stuff in there to keep it warm, or cool stuff to keep it cool. Thermos bottles have two walls: an inner and an outer wall. In between the two walls is an insulator. A vacuum is a really good insulator because it has no molecules or atoms of gas in it to conduct heat between the inner and outer walls of the thermos. Of course there will still be heat conduction along the walls. Let's think about a planet. There are no walls, however thin, connecting a planet to anything. The space between the Earth and the Sun is a pretty good vacuum. We know how warm it is in the sunshine, so we know that heat flows from the Sun to the Earth. Yet, separating the Sun from the Earth is 150 million km of vacuum. How can heat be carried through a vacuum?

Light carries heat through a vacuum. Electrons and protons have a property called ***electric charge***. What an electric charge is, fundamentally, no one can tell you, but it interacts through space via a property of the vacuum called the ***electric field***. A positive electric field attracts a negatively charged electron. This is how, in the olden days, a TV tube hurled electrons toward the screen in a picture-tube television until electronic flatscreens came along. It functioned by releasing an electron and then pushing it around through the electric field. The electric field interacts with another property of the vacuum called the ***magnetic field***.

If the strength of the electric field at some location, measured in volts, changes with time, and if the voltage oscillates up and down for example, this will cause a change in the magnetic field, such as a compass would point to. This is how an electromagnet works, converting electrical field energy into magnetic field energy. Going the other direction, if the magnetic field changes, it can produce an electrical field. This is how a generator works.

It turns out that the electric and magnetic fields in a vacuum fit together to form a closed cycle like the ringing of a bell. The up-and-down oscillation in an electric field will cause a complementary oscillation in the magnetic field, which reinforces the electric field in turn. The two fields "ring" together. Such a little bundle of electric and magnetic waves can in principle hurl through the vacuum forever, carrying energy with it.

The ringing of the electromagnetic field in light differs from the ringing of a piano string, in that light can come in any frequency. ***Frequencies***, of oscillators or of light waves, have units of cycles per second (hertz, Hz) and are denoted by the Greek letter ν (pronounced "new"). It turns out that different frequencies of light travel at the same speed in a vacuum. Within some nonvacuum medium, such as air, water, or glass, different frequencies of light might vary in their speeds a little bit, which is how a prism separates white light into its component colors. But in a vacuum, all light travels at the same speed. The ***speed of light in a vacuum***, c, is a fundamental constant of nature. The constancy of the speed of light in a vacuum makes it easy to relate the frequency of light to its ***wavelength***, the distance between the crests of a wave. We can figure out what the relationship is between frequency and wavelength by thinking geometrically, imagining the wavy line in Fig. 2.1 to be moving past us at some speed c. If the crests are 1 cm apart and moving at 10 cm/s, then 10 crests would move past us every second. Alternatively, we can make use of units. Pay attention to units, and they will lead you to virtue, or at least to the right answer. Problem: assemble the two things we know, ν and c, such that the units combine to be the same as the units of λ. Solution:

$$\lambda\left[\frac{\text{cm}}{\text{cycle}}\right] = \frac{c[\text{cm/s}]}{\nu[\text{cycle/s}]}$$

Don't take my word for it, check and make sure that the units are the same on both sides of the equation.

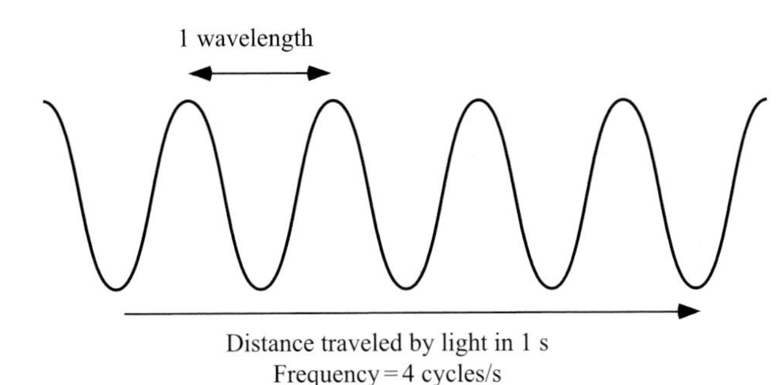

Fig. 2.1 **The frequency and wavelength of light are related to each other by the speed of light, which in a vacuum is the same for all different types of light, $3.0 \cdot 10^{10}$ cm/s.**

Scientists who discuss ***infrared*** (IR) light often use a third way of describing different colors, called the ***wave number***, which is defined as the number of cycles per centimeter of length. It's simply a matter of convenience; when IR light is described by its wave number, it will have a nice, easy-to-remember number to carry around in our memories. We will use the letter n to designate the wave number. How can we construct the units of n from the building blocks of λ, ν, and c? All we need is λ,

$$n\left[\frac{\text{cycle}}{\text{cm}}\right] = \frac{1}{\lambda[\text{cm/cycle}]}$$

Different frequencies of light all have the same fundamental nature; they are all waves of the same essential physics. Figure 2.2 shows the names assigned to different types of light based on their frequencies. Of course, if we know the frequency of light, we know its wavelength and wave numbers also, so we have added "milemarkers" of these units too.

Our eyes are sensitive to light in what we pragmatically call the visible range. Higher frequencies than this are in the ***ultraviolet***, or ***UV***, and x-ray ranges. UV light causes sunburn, and x-rays are even more dangerous. Objects that emit the highest frequency of light are considered radioactive. Extremely short wavelength light is called gamma radiation, and we might encounter it coming from processes that happen in the nuclei of atoms or processes that happen in space such as the explosions of stars, coming to us as cosmic rays. (Other common forms of "radiation" are made of flying electrons [beta radiation] or bundles of two each protons and neutrons, called alpha radiation.) At longer wavelengths than the visible range we find the IR range. Later we will find that objects at about room temperature glow with IR light. Heat lamps at the skating rink warm us by shining invisible IR light on us.

All that energy whizzing around space in the form of coupled electric and magnetic field waves would be of little interest to the energy balance of a planet if they did not give up or carry off energy. There are a number of different mechanisms by which light may interact with matter, but IR light interacts mostly with vibrations of the chemical bonds of a molecule. Light interacts with matter by means of the electric field that they share (Fig. 2.3). Let's imagine matter as constructed with charged oscillators all over its surface; little charged weights sticking out from the surface of the matter on

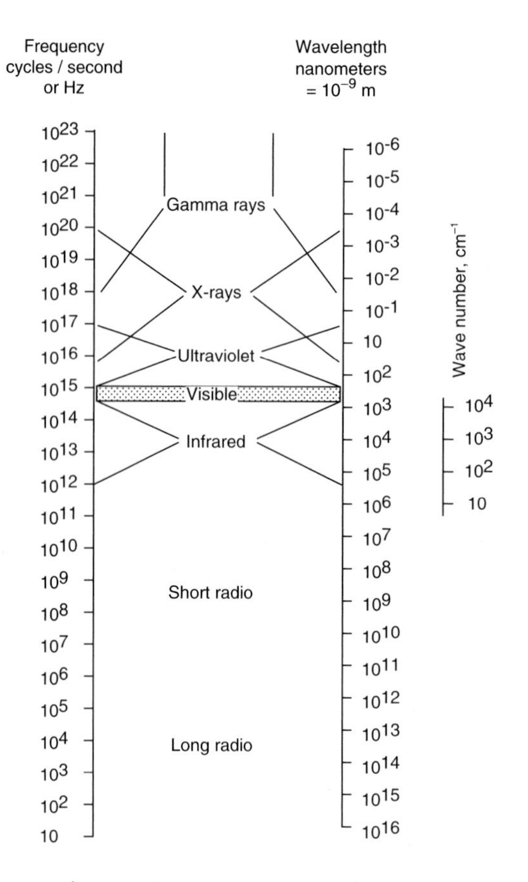

Fig. 2.2 **The electromagnetic spectrum.**

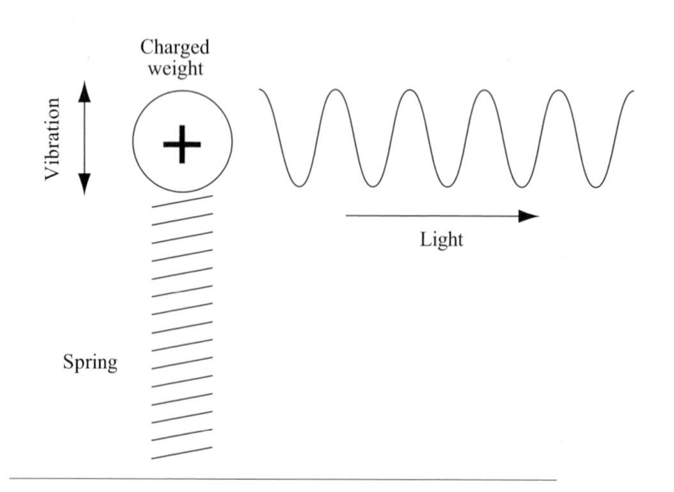

Fig. 2.3 **A charged oscillator interacting with light.**

little springs that stretch and contract. This little oscillator will have a frequency with which it will "ring." Incoming energy in the form of light brings with it an electric field oscillating up and down: voltage goes plus, minus, plus, minus. If the frequency of the cycles of voltage in the light is the same as the frequency of the oscillator light can be absorbed. Its energy is transferred into ***vibrational energy*** of the matter.

Important fact to remember: this mechanism of energy transfer is a two-way street. If energy can flow from the light to the oscillator, it can also flow the other way, from the oscillator to light. The vibrational energy of the oscillator is what we have been calling its temperature. Any matter that has a temperature above absolute zero (zero degrees on the Kelvin scale) will have energy in its oscillators that it may be able to use to create light. The two-way street character of this process is important enough that it is given the name of ***Kirchhoff's law***.

You can think of it as analogous to the vibrations of the oscillator strings of a piano interacting with the pressure field of the atmosphere to generate sound waves. You may have tried the experiment of singing into a piano with the dampers off to hear the strings echo back the note you sang into it. This wave-to-oscillator energy transfer is a two-way street as well.

Blackbody radiation

So where can we see electrical energy traveling the other way, from matter into light? One example: a red hot electric burner shines light you can see. The light derives its energy from the vibrations or thermal energy of the matter. We normally don't think of it, but it turns out that your electric burner continues to shine even when the stove is at room temperature. The difference is that the room temperature stove emits light in colors that we can't see, down in the IR range.

If we imagine our conceptual chunk of matter as having oscillators that vibrate at all possible frequencies, it would be able to absorb or emit all the different frequencies of light. We have a word for that: we call that object a ***blackbody***. The light that is emitted by a blackbody is called ***blackbody radiation***. Objects at room temperature, like the turned-off burner, emit light in the IR spectrum. The Sun acts as a blackbody but because it is so warm it emits visible light. And of course, if an object radiates at some frequency, it must also absorb at that frequency, so a blackbody is a perfect IR absorber as well.

Blackbody radiation is made up of a characteristic distribution of frequencies (colors)of IR light. Figure 2.4 shows a plot with axes of the intensity of light in the y-direction and frequency in the x-direction. The units of intensity look a bit complicated; they are W/m^2 wave number. The unit on the numerator is ***Watts*** (W), the same kind of Watts described by your hair dryers and audio amplifiers. A Watt is a rate of energy flow, defined as Joules per second, where energy is counted in ***Joules***. The meters squared on the denominator is the surface area of the ground. The unit of wave numbers on the denominator allow us to divide the energy up according to the different wave number bands of light; for instance, all of the light between 100 and 101 cm^{-1} carry this many Watts per square meter of energy flux, between 101 and 102 cm^{-1} carry that

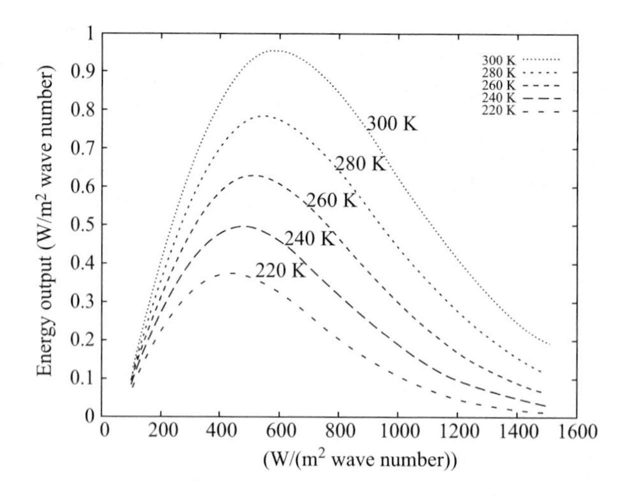

Fig. 2.4 **The intensity of light emitted from a blackbody as a function of light wave number, for different blackbody objects of different temperatures in Kelvins. A warmer object emits more radiation than a cooler one.**

many, and so on. We can calculate the total flux of energy by adding up the bits from all the different slices of the light spectrum. The plot is set up so that we can add up the area under the curve to obtain the total energy intensity in Watts per square meter. You could cut the plot out with a pair of scissors and weigh the inside piece to determine its area, which would give you the total energy emitted over all the wave numbers of light.

The intensity of light at each frequency is called a spectrum. The IR light emission spectrum of a blackbody depends only upon the temperature of the blackbody. There are two things you should notice about the shapes of the curves in Fig. 2.4.

First, as the temperature goes up, the curves are getting higher, meaning that light at each frequency is getting more intense (brighter). When we start talking about planets we will need to know how much energy is being radiated from a blackbody in total, over all wavelengths. The units in Fig. 2.4 were chosen specifically so that the total energy being carried by all frequencies of light is equal to the area under the curve of the spectrum. As the temperature of the object goes up, the total energy emitted by the object goes up, which you can see from the fact that the curves in Fig. 2.4 get bigger. There is an equation which tells us how quickly energy is radiated from an object. It is called the ***Stefan–Boltzmann equation***, and we are going to make extensive use of it. Get to know it now! The equation is

$$I = \varepsilon \sigma T^4 \tag{2.1}$$

The ***intensity*** of light is denoted by I, and it represents the rate of energy emission from the object. The Greek letter epsilon (ε) is the ***emissivity***, a number between zero and one describing how good the blackbody is. For a perfect blackbody, $\varepsilon = 1$. Sigma (σ) is a fundamental constant of physics which never changes, a number you can look up in reference books, called the ***Stefan–Boltzmann constant***. T is the temperature in Kelvins, and the superscript 4 is an exponent indicating that we have to raise the temperature to

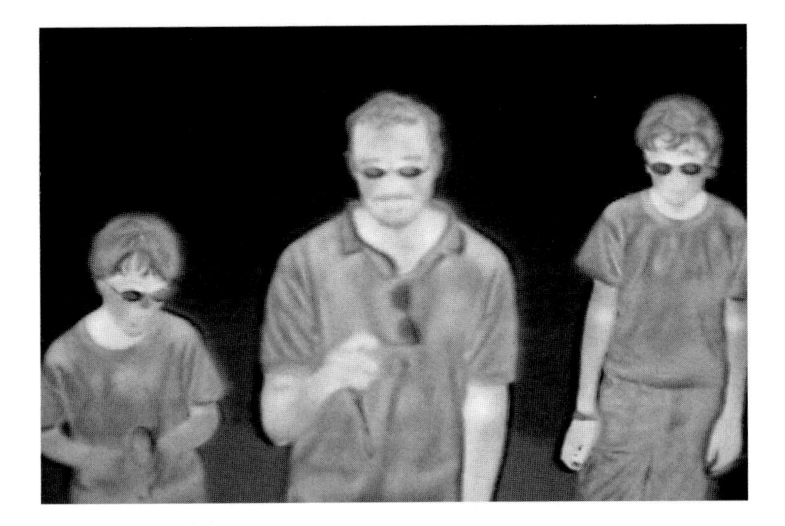

Fig. 2.5 **A photograph taken in IR light.**

the fourth power. The Kelvin temperature scale begins with 0 K when the atoms are vibrating as little as possible, a temperature called ***absolute zero***. There are no negative temperatures on the Kelvin scale.

One of the many tricks of thinking scientifically is to pay attention to units. Let us examine Eqn. (2.1) again, with units of the various terms specified in the brackets.

$$I\left[\frac{\mathrm{W}}{\mathrm{m}^2}\right] = \varepsilon[\text{unitless}]\ \sigma\left[\frac{\mathrm{W}}{\mathrm{m}^2\ \mathrm{K}^4}\right]\ T[\mathrm{K}]^4$$

The unit of energy flux is watts (W), equal to joules of energy per second. The meters squared on the bottom of that fraction is the surface area of the object that is radiating. The area of the earth, for example, is $5.14 \cdot 10^{14}\ \mathrm{m}^2$. Here's what I wanted to point out: the units on either side of this equation must be the same. On the right-hand side, K^4 cancels leaving only $\mathrm{W/m}^2$ to balance the left-hand side. In general, if are unsure about relating one number to another, the first thing to do is to listen to the units. We will see many more examples of units in our discussion, and you may rest assured I will not miss an opportunity to point them out.

IR-sensitive cameras allow us to see what the world looks like in IR light. The cheeks of the guy in Fig. 2.5 are warmer than the surface of his glasses, which are presumably at room temperature. We can estimate how much more light is shining from the cheeks than from the glass surface using Eqn. (2.1), to be

$$\frac{I_{\text{cheek}}}{I_{\text{glasses}}} = \frac{\varepsilon_{\text{cheek}}\sigma T^4_{\text{cheek}}}{\varepsilon_{\text{glasses}}\sigma T^4_{\text{glasses}}}$$

The Stefan–Boltzmann constant σ is the same on both top and bottom; σ never changes. The emissivity ε might be different between the cheek and the glasses, but let's assume

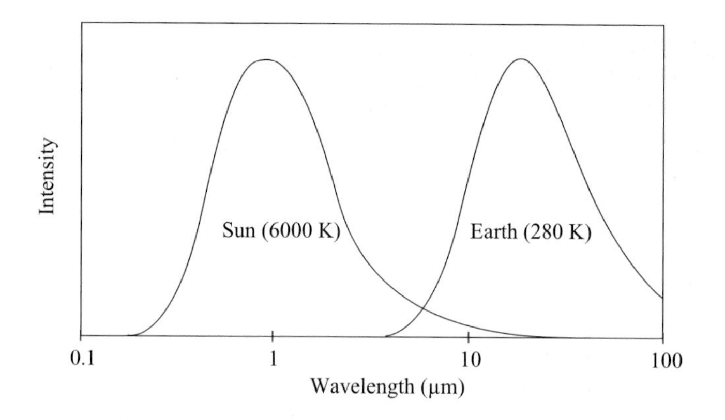

Fig. 2.6 **The shapes of the blackbody spectra of the Earth and the Sun. The Earth spectrum has been expanded to reach the same peak intensity as the solar spectrum, so the two can be compared on the same plot. The point is that the Sun shines in visible light while the Earth shines in IR light.**

that they are the same. This leaves us with the ratio of the brightnesses of the skin and glasses equal to the ratio of temperatures to the fourth power, maybe $(285\ \mathrm{K}/278\ \mathrm{K})^4$, which is about 1.1. The cheeks shine 10% more brightly than the surface of the coat, and that is what the IR camera shows us.

The second thing to notice about the effect of temperature on blackbody spectra in Fig. 2.4 is that the peaks shift to the right as temperature increases. This is the direction of higher frequency light. You already knew that a hotter object generates shorter wavelength light because you know about red hot and white hot. Which is hotter? White hot, of course; any kid on the playground knows that. An object at room temperature (say 273 K) glows in the IR range, which we can't see. A stove at stovetop temperatures (400–500 K) glows in shorter wavelength light, which begins to creep up into the visible part of the spectrum. The lowest energy part of the visible spectrum is red light. Get the object hotter, say the temperature of the surface of the Sun (5000 K), and it will fill all wavelengths of the visible part of the spectrum with light. Figure 2.6 compares the spectra of the Earth and the Sun. You can see that sunlight is visible whereas "earth light" (also referred to as ***terrestrial radiation***) is in the IR spectrum. Of course, the total energy flux from the Sun is much higher than it is from Earth. Repeating the calculation we used for the IR photo, we can calculate the ratio of the fluxes as $(5000\ \mathrm{K}/273\ \mathrm{K})^4$, or about 10^5. The two spectra in Fig. 2.6 have been scaled by dividing each curve by the maximum value that the curve reaches, so that the top of each peak is at a value of one. If we hadn't done that, the area under the Earth spectrum would be 100,000 times smaller than the area under the Sun spectrum, and you would need a microscope to see the Earth spectrum on Fig. 2.6.

It is not a coincidence that the Sun shines in what we refer to as visible light. Our eyes have evolved to be sensitive to visible light. The IR light field is a much more complicated thing for an organism to measure and understand. For one thing, the eyeball, or whatever light sensor the organism has, will be shining IR light of its own. The organism measures light intensity by measuring how intensely the incoming light deposits energy into oscillators coupled to its nervous system. It must complicate matters if the oscillators lose energy by radiating light of their own. IR telescopes must be cooled in order to make accurate IR intensity measurements. Snakes are capable of sensing IR light. Perhaps this is useful because their body temperatures are colder than those of their intended prey.

Take-home points

1. Light carries energy through space.
2. If an object can absorb light, it can also emit light.
3. An object that can emit all frequencies of light (a blackbody) emits light energy at a rate equal to σT^4.

Projects

1. A joule is an amount of energy, and a watt is a rate at which energy is used, defined as $1\,\text{W} = 1\,\text{J/s}$. How many joules of energy are required to run a 100 W light bulb for one day? Burning coal yields about $30 \cdot 10^6$ J of energy per kilogram of coal burned. Assuming that the coal power plant is 30% efficient, how much coal has to be burned to light the bulb for a day?

2. This is one of those job interview questions to see how creative you are, analogous to one I heard, "How many airplanes are over Chicago at any given time?." You need to make stuff up to get an estimate and demonstrate your management potential. The question is: What is the efficiency of energy production from growing corn?

Assume that sunlight deposits 250 W/m^2 of energy on a corn field, averaging over the day/night cycle. There are 4.186 J in a calorie. How many calories of energy are deposited on a square meter of field over the growing season? Now guess how many ears of corn grow per square meter, and guess what is the number of calories you get for eating an ear of corn. The word "calorie," when you see it on a food label, actually means "kilocalories," thousands of calories, so if you guess 100 food-label calories, you are actually guessing 100,000 true calories or 100 kcal. Compare sunlight energy with the corn energy to get efficiency.

3. The Hoover Dam produces $2 \cdot 10^9$ W of electricity. It is composed of $7 \cdot 10^9$ kg of concrete. To produce 1 kg of concrete requires 1 MJ of energy. How much energy did it take to produce the dam? How long is the "energy payback time" for the dam?

The area of Lake Mead, formed by Hoover Dam, is 247 mi^2. Assuming that 250 W/m^2 of sunlight falls on Lake Mead, how much energy could you produce if instead of the lake you installed solar cells that were 12% efficient?

4. It takes approximately $2 \cdot 10^9$ J of energy to manufacture 1 m^2 of crystalline silicon photovoltaic cell. (Actually the number quoted was 600 kWh. Can you figure out how to convert kilo-watt hours into joules?) Assume that the solar cell is 12% efficient, and

calculate how long it would take, given 250 W/m^2 of sunlight, for the solar cell to repay the energy it cost for its manufacture.

5. IR light has a wavelength of about 10 μm. What is its wave number (in cm^{-1})?
   Visible light has a wavelength of about 0.5 μm. What is its frequency (in Hz or cycles/s)?
   FM radio operates at a frequency of about 100 MHz. What is its wavelength?

Further reading

Blackbody radiation was a clue that something was wrong with classical physics, leading to the development of quantum mechanics. Classical mechanics predicted that an object would radiate an infinite amount of energy, instead of the $\varepsilon\sigma T^4$, as we observe it to be. The failure of classical mechanics is called the UV catastrophe, and you can read about it, lucidly but at a rather higher mathematical level, in the *Feynman Lectures on Physics*, vol. 1, chapter 41. My favorite book about quantum weirdness, the philosophical implications of quantum mechanics, is John Gribbon's *In Search of Schrodinger's Cat*, but there are many others.

3
The layer model

This is an algebraic calculation of the effect of an infrared (IR) absorber, a pane of glass essentially, on the mean temperature of the surface of the Earth. By solving the energy budgets of the Earth's surface and the pane of glass, the reader can see how the pane of glass traps outgoing IR light, leading to a warming of the surface. The layer model is not an accurate, detailed model suitable for a global warming forecast, but the principle of the greenhouse effect cannot be understood without understanding the layer model.

The bare rock model

The temperature of the surface of the Earth is controlled by the ways that energy comes in from the Sun and shines back out to space as IR. The Sun shines a lot of light because the temperature at the visible surface of the Sun is high and therefore the energy flux $I = \varepsilon\sigma T^4$ is a large number. Sunlight strikes the Earth and deposits some of its energy into the form of vibrations and other bouncings around of the molecules of the Earth. Neither the Earth nor the Sun is a perfect blackbody, but they are both almost blackbodies, as are most solids and liquids. (Gases are terrible blackbodies as we will learn in Chapter 4.) The Earth radiates heat to space in the form of IR light. Earth light is of much lower frequency and of lower energy than sunlight.

We are going to construct a simple model of the temperature of the Earth. The word ***model*** is used quite a bit in scientific discussions, to mean a fairly wide variety of things. Sometimes the word is synonymous with "theory" or "idea," such as the Standard Model of Particle Physics. For doctors, a "model system" might be a mouse that has some disease that resembles a disease that human patients get. They can experiment on the mouse rather than experiment on people. In climate science, models are used in two different ways. One way is to make forecasts. For this purpose, a model should be as realistic as possible and should capture or include all of the processes that might be relevant in Nature. This is typically a mathematical model implemented on a computer, although there's a nifty physical model of San Francisco Bay you should check out if you're ever in Sausalito. Once such a model has been constructed, a climate scientist can perform what-if experiments on it that could never be done on the real world, to determine how sensitive the climate would be to changes in the brightness of the Sun or properties of the atmosphere, for example.

The simple model that we are going to construct here is not intended for making predictions, but is rather intended to be a toy system that we can learn from. The model will demonstrate how the greenhouse effect works by stripping away lots of other aspects of the real world that would certainly be important for predicting climate change in the next century or the weather next week, but make the climate system more complicated and therefore more difficult to understand. The model we are going to explore is called the ***layer model***. Understanding the layer model will not equip us to make detailed forecasts of future climate, but one cannot understand the workings of the real climate system without first understanding the layer model.

The layer model makes a few assumptions. One is that the amount of energy coming into the planet from sunlight is equal to the amount of energy leaving the Earth as IR. The real world may be out of energy balance for a little while or over some region, but the layer model is always exactly in balance. We want to balance the energy budget by equating the ***outgoing energy flux*** F_{out} to the ***incoming energy flux*** F_{in},

$$F_{in} = F_{out}$$

Let's begin with incoming sunlight. The ***intensity of incoming sunlight*** I_{in} at the average distance from the Sun to the Earth is about 1350 W/m^2. We'll consider the Watts part of this quantity first, followed by the square meter part. If you've ever seen Venus glowing in the twilight sky you know that some of the incoming visible light shines back out again as visible light. Venus's brightness is not blackbody radiation; it is hot on Venus but not hot enough to shine white hot. This is ***reflected*** light. When light is reflected, its energy is not converted to vibrational energy of molecules in the Earth and then re-radiated according to the blackbody emission spectrum of the planet. It just bounces back out to space. For the purposes of the layer model, it is as if the energy had never arrived on Earth at all. The fraction of a planet's incoming visible light that is reflected back to space is called the planet's ***albedo*** and is given the symbol α (Greek letter alpha). Snow, ice, and clouds are very reflective and tend to increase a planet's albedo. The albedo of the bright Venus is high, 0.71, because of a thick layer of sulfuric acid clouds in the Venusian atmosphere, and is low, 0.15, for Mars because of lack of clouds on that planet. Earth's albedo of about 0.33 depends on cloudiness and sea ice cover, which might change with changing climate.

Incoming solar energy that is not reflected is assumed to be absorbed into vibrational energy of the molecules of the Earth. Using a present-day earthly albedo of 0.3, we can calculate that the intensity of sunlight that is absorbed by the Earth is $1350\,\mathrm{W/m^2}\,(1-\alpha) = 1000\,\mathrm{W/m^2}$.

What about the area, the square meters on the denominator? If we want to get the total incoming flux for the whole planet, in units of Watts instead of Watts per square meter, we need to multiply by a factor of area,

$$F_{in}[\mathrm{W}] = I\left[\frac{\mathrm{W}}{\mathrm{m}^2}\right] \cdot A[\mathrm{m}^2]$$

What area shall we use? Sun shines on half of the surface of the Earth at any one time, but the light is weak and wan on some parts of the Earth, during dawn or dusk or in high latitudes, but is much more intense near the Equator at noon. The difference in intensity is caused by the angle of the incoming sunlight, not because the sunlight,

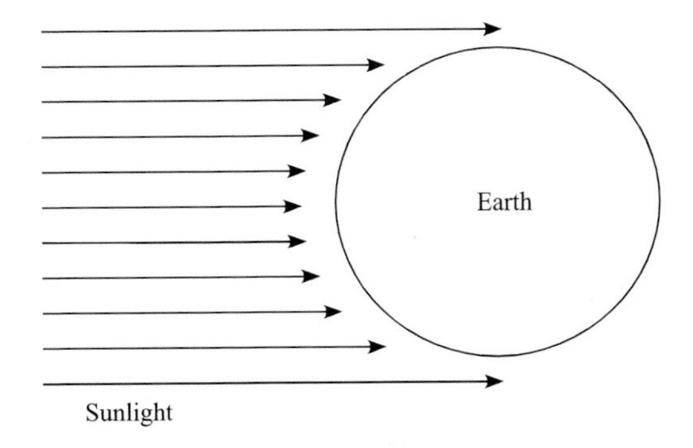

Fig. 3.1 **When sunlight hits the Earth, it all comes from the same direction. The Earth makes a circular shadow. Therefore, the Earth receives an influx of energy equal to the intensity of sunlight, multiplied by the area of the circle, $\pi \cdot r^2_{\text{earth}}$.**

measured head-on at the top of the atmosphere, is much different between low and high latitudes (Fig. 3.1). How then do we add up all the weak fluxes and the strong fluxes on the Earth to find the total amount of energy that the Earth is absorbing?

There's a nifty trick. Measure the size of the shadow. The area we are looking for is that of a circle, not a sphere. The area is

$$A[\text{m}^2] = \pi r^2_{\text{earth}}$$

Putting these together, the total incoming flux of energy to a planet by solar radiation is

$$F_{\text{in}} = \pi r^2_{\text{earth}}(1 - \alpha)I_{\text{in}}$$

Our first construction of the layer model will have no atmosphere, only a bare rock sphere in space. A real bare rock in space, such as the Moon or Mercury, is incredibly hot on the bright side and cold in the dark. The differences are much more extreme than they are on Earth or Venus where heat is carried by fluid atmospheres. Nevertheless, we are trying to find a single value for the temperature of the Earth, to go along with a single value for each of the heat fluxes F_{in} and F_{out}. The real world is not all at the same temperature, but we're going to ignore that in the layer model. The heat fluxes F_{in} and F_{out} may not balance each other in the real world, either, but they do in the layer model.

The rate at which the Earth radiates energy to space is given by the Stefan–Boltzmann equation:

$$F_{\text{out}} = A\varepsilon\sigma T^4_{\text{earth}}$$

As we did for solar energy, here we are converting intensity I to total energy flux F by multiplying by an area A. What area is appropriate this time? Incoming sunlight is different from outgoing earthlight in that the former travels in one direction whereas the latter leaves Earth in all directions (Fig. 3.2). Therefore, the area over which the Earth radiates energy to space is simply the area of the sphere, which is given by

$$A_{\text{sphere}} = 4\pi r^2_{\text{earth}}$$

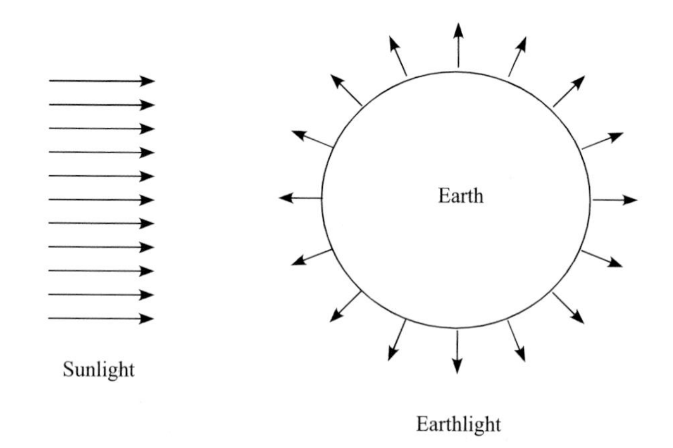

Fig. 3.2 **When IR light leaves the Earth, it does so in all directions. The total rate of heat loss equals the intensity of the earthlight multiplied by the area of the surface of the sphere $4 \cdot \pi \cdot r^2_{\text{earth}}$.**

Therefore, the total outgoing energy flux from a planet by blackbody radiation is

$$F_{\text{out}} = 4\pi r^2_{\text{earth}} \varepsilon \sigma T^4_{\text{earth}}$$

The layer model assumes that the energy fluxes in and out balance each other (Fig. 3.3)

$$F_{\text{out}} = F_{\text{in}}$$

which means that we can construct an equation from the "pieces" of F_{out} and F_{in} which looks like this:

$$4\pi r^2_{\text{earth}} \varepsilon \sigma T^4_{\text{earth}} = \pi r^2_{\text{earth}} (1 - \alpha) I_{\text{in}}$$

Factors of π and r_{earth} appear in common on both sides of the equation, which means that we can cancel them by dividing both sides of the equation by those factors. Also dividing by 4 on both sides gives units of Watts per area of the Earth's surface. We get

$$\varepsilon \sigma T^4_{\text{earth}} = \frac{(1 - \alpha) I_{\text{in}}}{4}$$

We know everything here except the T_{earth}. If we rearrange the equation to put what we know on the right-hand side and what we don't on the left, we get

$$T_{\text{earth}} = \sqrt[4]{\frac{(1 - \alpha) I_{\text{in}}}{4 \varepsilon \sigma}} \tag{3.1}$$

What we have constructed is a relationship between a number of crucial climate quantities. Changes in solar intensity such as the sunspot cycle or the Maunder Minimum (Chapter 10) may affect I_{in}. We shall see in Chapter 4 that greenhouse gases

Table 3.1 The temperatures and albedos of the terrestrial planets. The intensity of sunlight differs with distance from the Sun

	I_{solar} (W/m^2)	α (%)	T_{bare} (K)	$T_{observed}$ (K)	$T_{1\ layer}$ (K)
Venus	2600	71	240	700	285
Earth	1350	33	251	295	303
Mars	600	17	216	240	259

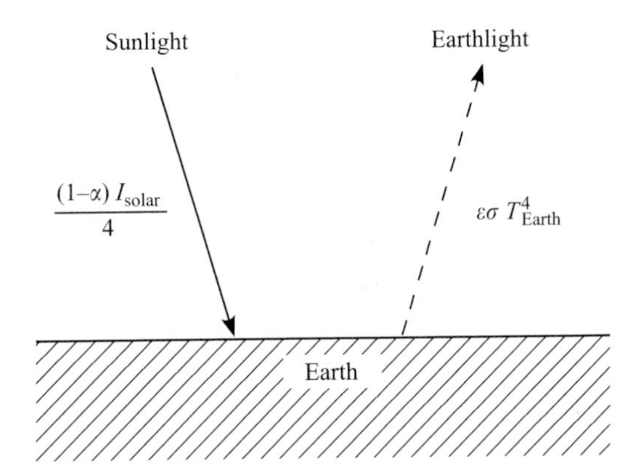

Fig. 3.3 **An energy diagram for Earth with no atmosphere, just a bare rock in space.**

are extremely selective about the wavelengths of light that they absorb and emit; in other words, they have complexities relating to their emissivity ε values. The albedo of the planet is very sensitive to ice and cloud cover, both of which might change with changing climate.

If we calculate the temperature of the Earth, we get a value of 255 K or about −15°C. This is too cold; the temperature range of Earth's climate is −80°C to about +55°C, but the average temperature, what we're calculating using the layer model, is closer to +15°C than −15°C. Table 3.1 gives the values we need to do the same calculation for Venus and Mars, along with the results of the calculation and the observed average temperatures. In all three cases, our calculation has erred on the side of too cold.

The layer model with greenhouse effect

Our simple model is too cold because it lacks the ***greenhouse effect***. We had no atmosphere on our planet; what we calculated was the temperature of a bare rock in space (Fig. 3.3), like the Moon. Of course the surface of the moon has a very different temperature on the sunlit side than it does in the shade, but if the incoming sunlight were spread out uniformly over the Moon's surface, or if somehow the heat from one side

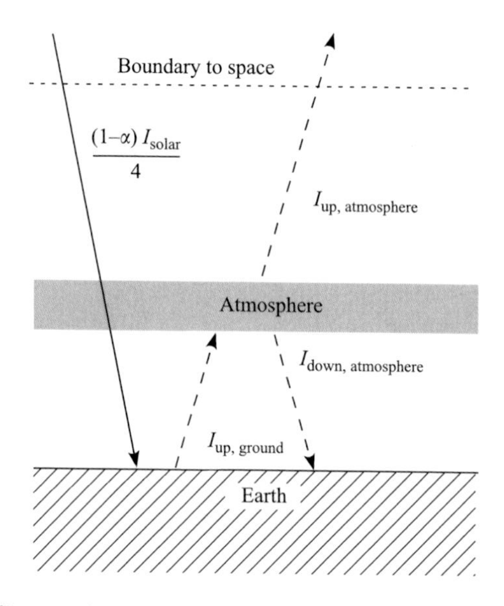

Fig. 3.4 **An energy diagram for a planet with a single pane of glass for an atmosphere. The glass is transparent to incoming visible light, but a blackbody to IR light.**

of the planet conducted quickly to the other side, or if the planet rotated real fast, our calculation would be pretty good. But to get the temperatures of Earth, Venus, and Mars, we need a greenhouse effect.

In keeping with the philosophy of the layer model, the atmosphere in the layer model is simple to the point of absurdity. The atmosphere in the layer model resembles a pane of glass suspended by magic above the ground (Fig. 3.4). Like glass, our atmosphere is transparent to visible light, so the incoming energy from the Sun passes right through the atmosphere and is deposited on the planet surface, as before. The planet radiates energy as IR light according to $\varepsilon\sigma T_{\text{ground}}$, as before. In the IR range of light, we will assume that the atmosphere, like a pane of glass, acts as a blackbody, capable of absorbing and emiting all frequencies of IR light. Therefore the energy flowing upward from the ground, in units of watts per square meter of the Earth's surface, which we will call $I_{\text{up, ground}}$, is entirely absorbed by the atmospheric pane of glass. The atmosphere in turn radiates energy according to $\varepsilon\sigma T_{\text{atmosphere}}$. Because the pane of glass has two sides, a top side and a bottom side, it radiates energy both upward and downward, $I_{\text{up, atmosphere}}$ and $I_{\text{down, atmosphere}}$.

The layer model assumes that the energy budget is in steady state; energy in = energy out. This is true of any piece of the model, such as the atmosphere, just as it is for the planet as a whole. Therefore, we can write an ***energy budget for the atmosphere***, in units of watts per area of the earth's surface, as

$$I_{\text{up, atmosphere}} + I_{\text{down, atmosphere}} = I_{\text{up, ground}}$$

or

$$2\varepsilon\sigma T^4_{\text{atmosphere}} = \varepsilon\sigma T^4_{\text{ground}}$$

The budget for the ground is different from before because we now have heat flowing down from the atmosphere. The basic balance is

$$I_{out} = I_{in}$$

We can break these down into component fluxes

$$I_{up,\,ground} = I_{in,\,solar} + I_{down,\,atmosphere}$$

and then further dissect them into

$$\varepsilon\sigma T_{ground}^4 = \frac{(1-\alpha)}{4} I_{solar} + \varepsilon\sigma T_{atmosphere}^4$$

Finally, we can also write a ***budget for the earth overall*** by drawing a boundary above the atmosphere and figuring that if energy gets across this line in, it must also be flowing across the line out at the same rate.

$$I_{up,\,atmosphere} = I_{in,\,solar}$$

The intensities are comprised of individual fluxes from the Sun and from the atmosphere

$$\varepsilon\sigma T_{atmosphere}^4 = \frac{(1-\alpha)}{4} I_{solar}$$

There is a solution to the layer model for which all the budgets balance. We are looking for a pair of temperatures T_{ground} and $T_{atmosphere}$. Solving for T_{ground} and $T_{atmosphere}$ is a somewhat more complex problem algebraically than the bare rock model with no atmosphere we solved above, but we can still do it. We have two unknowns, and we appear to have three equations, the budgets for the atmosphere, for the ground, and for the Earth overall. Three equations and two unknowns might be a recipe for an unsolvable system,but it turns out that in this problem we are free to use any two of the three budget equations to solve for the unknowns T_{ground} and $T_{atmosphere}$. The third equation is simply a combination of information from the first two. The budget equation for the Earth overall, for example, is just the sum of the budget equations for the ground and the atmosphere (verify this for yourself).

There are laborious ways of approaching this problem, and there is also an easy way. Shall we choose the easy way? OK. The easy way is to begin with the energy budget for the Earth overall. This equation contains only one unknown, $T_{atmosphere}$. Come to think of it, this equation looks a lot like Eqn. (3.1), describing the surface temperature of the bare planet model above. If we solve for $T_{atmosphere}$ here, we get the same answer as when we solved for $T_{bare\ earth}$. This is an important point, more than just a curiosity or an algebraic convenience. It tells us that the place in the Earth system where the temperature is most directly controlled by the rate of incoming solar energy is the temperature at the location that radiates to space. We will call this temperature the ***skin temperature*** of the Earth.

What about temperatures below the skin, in this case T_{ground}? Now that we know that the outermost temperature, $T_{\text{atmosphere}}$, is equal to the skin temperature, we can plug that into the budget equation for the atmosphere to see that

$$2\varepsilon\sigma T_{\text{atmosphere}}^4 = \varepsilon\sigma T_{\text{ground}}^4$$

or

$$T_{\text{ground}} = \sqrt[4]{2}\, T_{\text{atmosphere}}$$

The temperature of the ground must be warmer than the skin temperature, by a factor of the fourth root of two, an irrational number that equals about 1.189. The ground is warmer than the atmosphere by about 19%. When we do the calculation T_{ground} for Venus, Earth, and Mars in Table 3.1, we see that we are getting Earth about right, Mars too warm, and Venus not yet warm enough.

The blackbody atmospheric layer is not a source of energy, like some humungous heat lamp in the sky. How then does it change the temperature of the ground? I am going to share with you what is perhaps my favorite earth sciences analogy, that of the equilibrium water level in a steadily filled and continuously draining sink. Water flowing into the sink, residing in the sink for a while, and draining away is analogous to energy flowing into and out of the planet. Water drains faster as the level in the sink rises, as the pressure from the column of water pushes water down the drain. This is analogous to energy flowing away faster as the temperature of the planet increases, according to $\varepsilon\sigma T^4$. Eventually the water in the sink reaches a level where the outflow of water balances the inflow. That's the equilibrium value and is analogous to the equilibrium temperature we calculated for the layer model. We constrict the drain somewhat by putting a penny down on the filter. For a while, the water drains out slowly, and the water level in the sink rises because of the water budget imbalance. The water level rises until the higher water level pushes water down the drain fast enough to balance the faucet again. A greenhouse gas, like the penny in the drain filter, makes it more difficult for the heat to escape the Earth. The temperature of the Earth rises until the fluxes balance again.

Take-home points

1. The outflow of IR energy from a planet must balance heating from the Sun.
2. The planet accomplishes this act of energetic housekeeping by adjusting its temperature.
3. Absorption of outgoing IR light by the atmosphere warms the surface of the planet, as the planet strives to balance its energy budget.

Projects

1. *The moon with no heat transport.* The layer model assumes that the temperature of the body in space is all the same. This isn't really very accurate, as you know that it's colder at the poles than it is at the equator. For a bare rock with no atmosphere or ocean, like the Moon, the situation is even worse because fluids like air and water are how heat is carried around on the planet. So let's make the other extreme assumption, that there is no heat transport on a bare rock like the Moon. What is the equilibrium temperature of the surface of the Moon, on the equator, at local noon, when the Sun is directly overhead? What is the equilibrium temperature on the dark side of the Moon?

2. *A two-layer model.* Insert another atmospheric layer into the model, just like the first one. The layer is transparent to visible light but a blackbody for IR

a. Write the energy budgets for both atmospheric layers, for the ground, and for the Earth as a whole, just like we did for the one-layer model.
b. Manipulate the budget for the Earth as a whole to obtain the temperature T_2 of the top atmospheric layer, labeled atmospheric layer 2 in Fig. 3.5. Does this part of the exercise seem familiar in any way? Does the term skin temperature ring any bells?
c. Insert the value you found for T_2 into the energy budget for layer 2, and solve for the temperature of layer 1 in terms of layer 2. How much bigger is T_1 than T_2?
d. Now insert the value you found for T_1 into the budget for atmospheric layer 1, to obtain the temperature of the ground, T_{ground}. Is the greenhouse effect stronger or weaker because of the second layer?

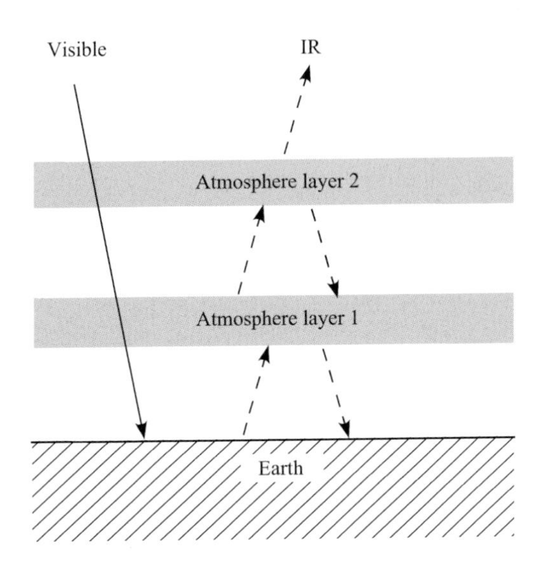

Fig. 3.5 **An energy diagram for a planet with two panes of glass for an atmosphere.**

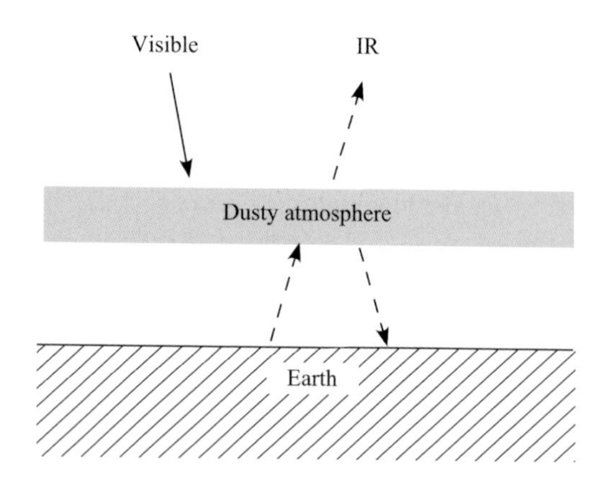

Fig. 3.6 **An energy diagram for a planet with an opaque pane of glass for an atmosphere.**

3. *Nuclear winter.* Let's go back to the one-layer model, but let's change it so that the atmospheric layer absorbs visible light rather than allowing to pass through (Fig. 3.6). This could happen if the upper atmosphere were filled with dust. For simplicity, let's assume that the albedo of the Earth remains the same, even though in the real world it might change with a dusty atmosphere. What is the temperature of the ground in this case?

Further reading

Kasting, J.F. and D. Catling, Evolution of a habitable planet, *Annual Review of Astronomy and Astrophysics* (2003), 41, 429–63.

Kump, Lee R., James F. Kasting, and Robert G. Crane, *The Earth System*, Prentice-Hall, 1999.

4
Greenhouse gases

The layer model assumes that the atmosphere acts as a blackbody in the infrared (IR), absorbing and emitting all frequencies of IR light. In reality, gases absorb IR light selectively, and most of the gases in the atmosphere do not interact with IR light at all. The difference can be understood in terms of the effect of molecular vibration on the electromagnetic field. Because gases absorb IR selectively, there are some radiation bands that are completely absorbed (the gases are saturated), and others such as the atmospheric window, where no gases absorb. This leads to much higher greenhouse forcing per molecule from some trace gases, such as freons, SF_6, or to a lesser extent methane, than from more abundant gases such as CO_2. Some absorption bands fall in the middle of the IR emission spectrum of the Earth's surface, while other bands fall outside this spectrum and are therefore irrelevant to the heat budget.

About gases

The layer model is what we call an idealization of the real world. Now that we understand the core mechanism of the greenhouse effect, by understanding the layer model, we can add things one at a time from the real world, and see how they affect the way that Earth's temperature is controlled. The first modification we have to make to the layer model is to think more about real gases in the atmosphere.

Let's begin by defining different ways of describing the amounts of gases in the atmosphere. The word ***concentration*** means the number of molecules within some volume. The difficulty this raises for gases in the atmosphere is that there are fewer molecules per volume overall as the gas expands. The major gases in the atmosphere are pretty well mixed so that the concentrations of these gases go down as the gas expands with altitude. This is why it's hard to breathe oxygen quickly enough on Mount Everest. It is often more convenient to talk about proportions of gases, like oxygen is about 20% of the molecules of gas in the atmosphere, and nitrogen almost 80%. The proportion of CO_2 is currently 0.038%. We can express this in a more convenient way by saying 380 parts per million (ppm). This number is called a ***mixing ratio***. The mixing ratio of a gas is numerically equal to the pressure exerted by the gas, denoted for CO_2 as ***p***$\boldsymbol{CO_2}$. In 2005 as I write this, the pCO_2 of the atmosphere is reaching 380 μatm. It is rising by about 1.5 ppm/year, which is the same as saying 1.5 μatm/year.

Gases, vibrations, and light

Most of the mass of an atom is in its nucleus, which resembles the massive Sun at the center of the solar system. Electrons float in ghostly quantum mechanical probability clouds, called ***orbitals***, around the nucleus. The two nuclei of two different atoms always repel each other because of their positive charges. The orbitals for the electrons fit together better, however, with certain numbers of orbitals than with others. The electrons from two different atoms may be able to combine their orbitals in such a way that they are of lower energy, as if happier, when they share a ***chemical bond***. A chemical bond is like a spring in that the two nuclei on either end of the bond have some freedom to move closer or farther apart. There is an optimum distance for the nuclei to be from each other. Closer, and the positive nuclei will start to repel each other. Farther, and you get less energy gain from sharing the electrons. A bond vibrates when the distance between the nuclei oscillates between the nuclei being too close together, then too far apart.

Gases are the simplest type of molecules, and they only vibrate in very particular ways. Vibrations in a gas molecule are like the vibrations of a piano string in that they are fussy about frequency. This is because, like a piano string, a gas molecule will only vibrate at its "ringing" frequency. The ringing frequency of an oscillator made of weights and springs depends on two things: the amount of weight on the ends and the strength of the spring holding them together. Heavy weights will have enough inertia to keep a bond growing in the wrong direction for longer than will a pair of light weights, so the frequency of the vibration will be slower. If the spring is very strong, it will reverse the velocity of a vibration more quickly, and the frequency of the oscillation will be higher. Vibrations in chemical bonds depend on the mass of the nuclei and on the energy penalty for having the nuclei too close or too far apart: the springiness of the chemical bond.

However, the vibrations of many gas molecules, such as the major gases in the atmosphere oxygen and nitrogen, are invisible to the electromagnetic field. They don't shine light or absorb IR light; we say they are not ***infrared active***. Oxygen and nitrogen are not greenhouse gases because they are transparent to infrared light. These molecules are invisible because when you stretch one, it doesn't change the electric field. These are symmetric molecules made of two identical atoms whose electric fields just cancel each other out. Neither atom can hold the electrons any more tightly than the other. In general, symmetrical molecules with only two atoms are not greenhouse gases.

We can break the symmetry by making a molecule of NO for example. This is a very reactive molecule, an ingredient for producing urban smog, but that's another story. NO has one atom of each element and, as a result, has a slight imbalance in its distribution of electrons. One side of the molecule will have a slight positive charge, and the other a slight negative charge. We could oscillate the electric field simply by rotating an NO molecule. Also, if we vibrate an NO molecule, the steepness of the transition from slightly positive to slightly negative will oscillate with time. By these mechanisms, NO could be a greenhouse gas, but it turns out not to be a very important one because there is not very much of it.

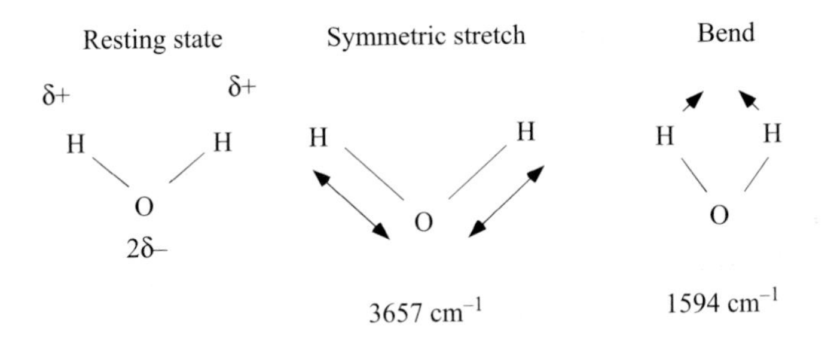

Fig. 4.1 **Vibrational modes of a water molecule that interact with IR light in the atmosphere.**

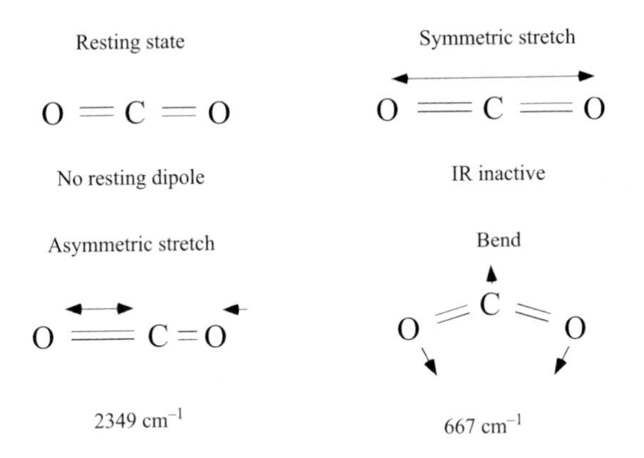

Fig. 4.2 **Vibrational modes of a CO_2 molecule that interact with IR light in the atmosphere.**

Molecules with more than two atoms have more than one chemical bond. All of their bonds ring together rather than each bond ringing with its own characteristic frequency. Water, H_2O, is a molecule that is bent in its lowest energy state (Fig. 4.1). This is because several of the electron orbitals stick off in the direction that appears in my diagram to be empty space. Hydrogen atoms hold their electrons more loosely than oxygen atoms, and so each hydrogen has a slightly positive charge (marked in Fig. 4.1 using the lowercase Greek letter delta, as $\delta+$). The oxygen end of the molecule has a slight negative charge. Just as for the NO molecule, rotating a H_2O molecule would oscillate the electric field and generate light. Because the arrangement of the nuclei in H_2O is more complex than for NO, there are several modes of vibration of the water molecule, including a symmetric stretch and a bend. These modes are also IR active.

The CO_2 molecule is shaped in a straight line with carbon in the middle (Fig. 4.2). It is a symmetric molecule; the oxygen atom on one end pulls the electrons just as tightly as the other oxygen on the other end. Therefore, rotating the molecule at rest has no effect on the electric field. Nor does a symmetric stretch. However, there are two modes of vibration which do generate an asymmetry in the electric field. One is an asymmetric stretch, and the other is a bend. The bend is the most climatically important one, as we shall see next.

How a greenhouse gas interacts with earthlight

We have seen that gases are terrible blackbodies because they are very choosy about which frequencies they absorb and emit. What we will now see is that some frequency bands are more important to the climate of the Earth than others. There are two factors to consider. One is the concentration of the gas, which we will discuss below. The other is the frequency of the absorption band relative to the blackbody spectrum for the Earth.

Figure 4.3 shows blackbody spectra again for temperatures ranging from 300 K, a hot summer day, down to 220 K, which is about the coldest it gets in the atmosphere, up near the troposphere at about 10 km altitude. There is also a jagged-looking curve. This is the intensity of light that an IR spectrometer would see if it were in orbit over the Earth, looking down. Figure 4.3 is not data, but rather a model simulation from one of our online models. You can point a web browser at http://understandingtheforecast.org/models/infrared_spectrum.html to run this model yourself. We will do so in the projects at the end of this chapter.

The spectrum of light leaving the Earth going into space ranges between two different blackbody spectra, a warmer one of about 270 K and a colder one from about 220 K.

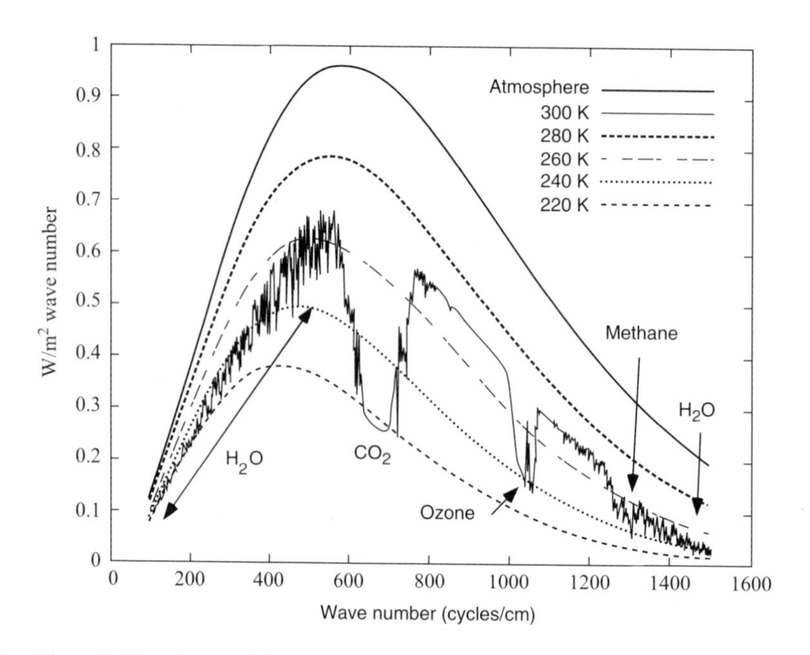

Fig. 4.3 **The solid line is a model-generated spectrum of the IR light escaping to space at the top of the atmosphere. For comparison, the broken lines are blackbody spectra at different temperatures. If the Earth had no atmosphere, the outgoing spectrum would look like a blackbody spectrum for 270 K, between the 260 and 280 K spectra shown. The atmospheric window is between about 900 and 1000 cm^{-1}, where no gases absorb or emit IR light. CO_2, water vapor, ozone, and methane absorb IR light emitted from the ground and emit lower-intensity IR from high altitudes where the air is colder than at the surface.**

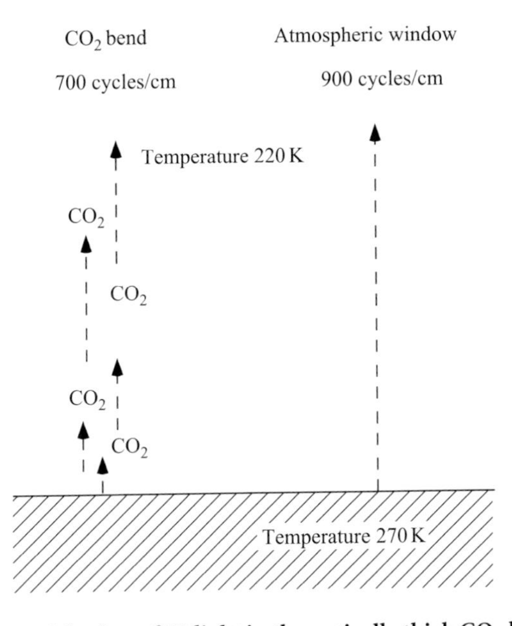

Fig. 4.4 **A comparison of the fate of IR light in the optically thick CO_2 bend frequency (left) versus the optically thin atmospheric window (right).**

The parts of the spectra that seem to follow the colder blackbody curve come from greenhouse gases in the upper atmosphere. They follow the colder blackbody curve because it is cold in the upper atmosphere. The most pronounced of these absorption bands, centered on a wave number of about 700 cycles/cm, comes from the bending vibration of CO_2. Light of this intensity that shines from the surface of the Earth is absorbed by the CO_2 in the atmosphere (Fig. 4.4). The CO_2 in the atmosphere then radiates its own light at this frequency. Remember from Chapter 1 that light emission and absorption is a two-way street.

Other parts of the spectrum, most notably the broad smooth part around 1000 cycles/cm, follow a warmer blackbody spectrum. These come directly from the ground. The atmosphere is transparent to IR light in these frequencies. This band is called the ***atmospheric window***.

The situation is analogous to standing on a pier and looking down into a pond of water. If the water were very clear, you could see light coming from the bottom; you would see rocks or old tires or whatever in the reflected light. If the water were murky, the light you would see would be scattered light coming from perhaps just a few inches down into the water. The old tires would be invisible, alas.

Remember we said that the total energy flux from one of these spectra can be "eye-balled" as the total area under the curve. The areas of the pure blackbody curves are going up proportionally to the temperature raised to the fourth power because of the Stefan–Boltzmann equation (Eqn. 2.1 in Chapter 2). The area trick works with our new jagged spectrum as well. The effect of an atmospheric absorption band is to take a bite out of the blackbody spectrum from the Earth's surface, decreasing the area and therefore decreasing the outgoing energy flux.

Compare the CO_2 absorption band at 700 cycles/cm with the absorption band of methane at around 1300 cycles/cm. The CO_2 band has a lot more room to change the outgoing IR energy flux than does the methane band, simply because the Earth and the atmosphere radiate a lot more energy near 700 cycles/cm than near 1300 cycles/cm. Both blackbody spectra are of pretty low intensity in the methane band.

Band saturation

The core of the CO_2 bend absorption band, between 600 and 800 cycles/cm, looks smooth rather than jagged and it follows a blackbody spectrum from about 220 K. This is about as cold as the atmosphere gets, and if we change the amount of CO_2 in the atmosphere, the intensity of light in this range does not get any lower (Fig. 4.5). We call this phenomenon ***band saturation***. You can see it in a series of model runs in which the CO_2 concentration of the atmosphere goes up from 0 to 1000 ppm. The current concentration of CO_2 in the atmosphere is about 380 ppm, as we will learn more in Part II. If there were no CO_2 in the atmosphere, the atmosphere would be

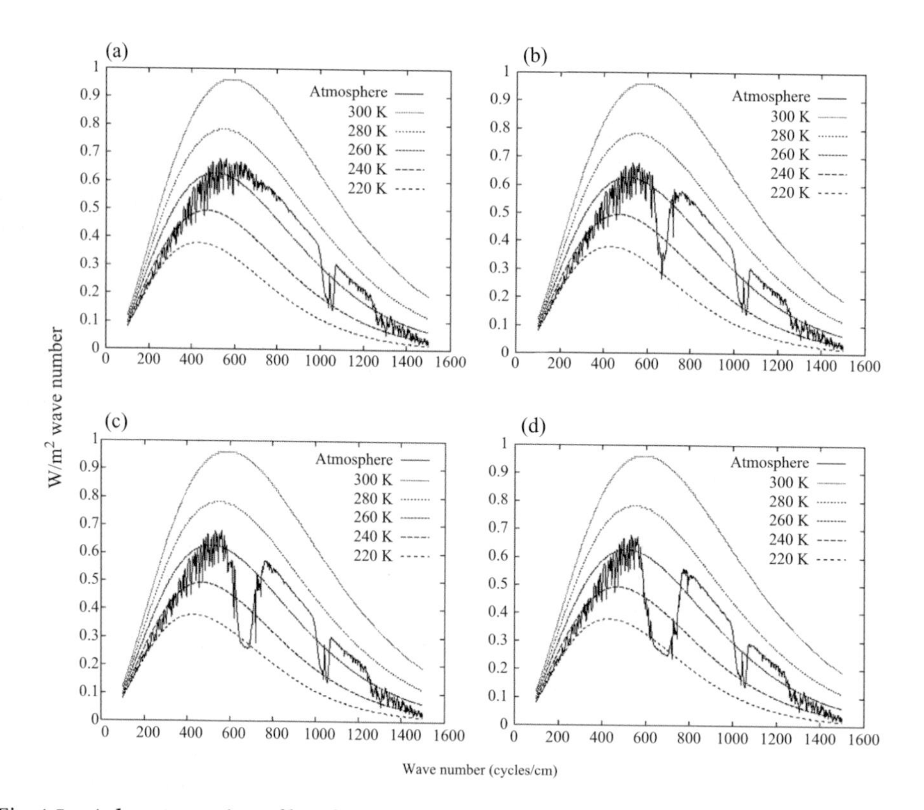

Fig. 4.5 **A demonstration of band saturation by CO_2. The addition of (b) 10 ppm CO_2 makes a huge difference to the outgoing IR light spectrum relative to an atmosphere that has (a) no CO_2. Increasing CO_2 to (c) 100 and (d) 1000 ppm continues to affect the spectrum, but you get less bang for your CO_2 buck as CO_2 concentration gets higher.**

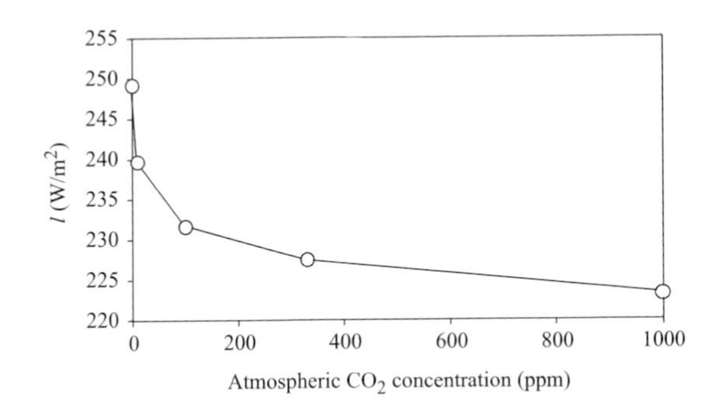

Fig. 4.6 **Band saturation viewed in a different way from Fig. 4.5. This is a plot of the total energy flux carried by all IR light, which is proportional to the area under the spectrum curves in Fig. 4.5. The outgoing energy flux is less sensitive to CO_2 when CO_2 concentration is high.**

transparent to light of around 700 cycles/cm, as it is in the atmospheric window. Adding the first 10 ppm of CO_2 has a fairly noticeable impact on the shape of the outgoing light spectrum, but increasing CO_2 from say 100 to 1000 ppm has a somewhat subtler effect.

I have plotted the total energy intensity I_{out} (in W/m^2) as a function of the concentration of CO_2 in the atmosphere (Fig. 4.6). Changes in CO_2 concentration have the greatest effect if we were starting out from no CO_2 and adding just a bit. The first 10 ppm of added CO_2 changes I_{out} by as much as going from 10 to 100, or 100 to 1000 ppm. We can understand why by analogy to our murky pond or by looking back at Fig. 4.4. As we increase the murkiness of the water, we decrease the distance that a photon of light can travel before it is absorbed. It doesn't take much murk in the water to obscure the old tire on the bottom, shifting the depth to which we can see from the bottom at say 3 m to maybe only 1 m. If we make the pond a lot murkier we will only be able to see a few centimeters down into the water. Making it murkier still will limit our view to only 1 cm. The change in depth is getting less sensitive to the murkiness of the pond. In the same way, the changes in the temperature at which the atmosphere radiates to space get smaller as the CO_2 concentration of the air gets higher. You just see the coldest light that you can get.

The band saturation for CO_2 makes CO_2 a less potent greenhouse gas than it would be if we had no CO_2 in the air to start with. Let's revisit our comparison of the CO_2 and methane as greenhouse gases. Methane had a disadvantage because its absorption band sort of fell in the suburbs of the earthlight spectrum whereas CO_2 fell right downtown. Now we see the advantage shifting the other way. Methane has a much lower concentration in the atmosphere. You can see from the jagged edges of the methane peak in Fig. 4.3 that the methane absorption band is not saturated. For this reason, in spite of the suburban location of the methane band, a ***molecule of methane*** added to the atmosphere is ***20 times more powerful than is a molecule of CO_2***.

If the edges of the absorption bands were completely abrupt, as if CO_2 absorbed 600 cycles/cm light completely and 599 cycles/cm light not at all, then once an absorption band from a gas was saturated, that would be it. Further increases in the concentration of the gas would have no impact on the radiation energy budget for the Earth.

CO_2, the most saturated of the greenhouse gases, would stop changing climate after it exceeded some concentration. It turns out that this is not how it works. Even though the core of the CO_2 band is saturated, the edges of the band are not saturated. When we increase the CO_2 concentration, the bite that CO_2 takes out of the spectrum doesn't get deeper, but it gets a bit broader.

The bottom line is that the energy intensity I_{out} in units of Watts per square meter goes up proportionally to the log of the CO_2 concentration, rather than proportionally to the CO_2 concentration itself (we would say linear in CO_2 concentration). The logarithmic dependence means that you get the same I_{out} change in Watts per square meter from any doubling of the CO_2 concentration. The radiative effect of going from 10 to 20 μatm pCO_2 is the same as going from 100 to 200 μatm, or 1000 to 2000 μatm.

The sensitivities of climate models are often compared as the average equilibrium temperature change from doubling CO_2, a diagnostic number that is called ΔT_{2x}. Most models have a ΔT_{2x} between 2 and 5 K, which is the same as between 2°C to 5°C. You can use ΔT_{2x} to estimate a temperature change resulting from some change in CO_2. Note that this is the ultimate temperature change, after hundreds or even thousands of years have passed (see Chapters 7 and 12). The equation is

$$\Delta T = \Delta T_{2x} \times \frac{\ln(\text{new } pCO_2/\text{orig} \cdot pCO_2)}{\ln(2)} \tag{4.1}$$

where ln is the natural log, the reverse operation of the exponential function e^x, The symbol "e" denotes a number which has no name other than simply "e". We will meet "e" again in Chapter 5. The exponential function is to raise "e" to the power of x. If

$$e^x = y \tag{4.2}$$

then

$$y = \ln(y) \tag{4.3}$$

Equilibrium temperature changes from changes in CO_2, assuming various ΔT_{2x} values, are shown in Fig. 4.7.

What happens to the energy balance of the Earth if we add a greenhouse gas to its atmosphere? If the energy budget was in equilibrium before, it isn't any more because the greenhouse gas has decreased the amount of energy leaving the Earth to space. We can see this visually as the big bite out of the spectrum going from the top to the middle diagram in Fig. 4.8. The decrease in energy flux is proportional to the area of that bite, the difference between (a) and (b) in Fig. 4.8. Referring to Chapter 2, the premise of the layer model is that the energy coming into and going out of the planet must balance, and the planet accomplishes this feat by adjusting its temperature. If we want to rebalance the energy flux after kicking it by adding CO_2, we do that by increasing the temperature of the ground. Using the online model, we find that a temperature change of 8.5 K brings us back to the same energy output I_{out} as we had before. Looking at Fig. 4.8(c), we see that the new, warmer output spectrum has risen

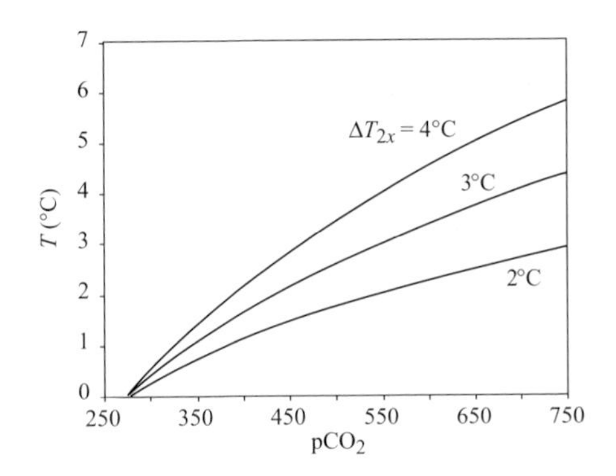

Fig. 4.7 **The average temperature of the Earth as a function of atmospheric CO_2 concentration and the climate sensitivity parameter, ΔT_{2x}.**

everywhere compared to (b). Visually, we have cut some area out of the CO_2 absorption band, and added it in the atmospheric window and other parts of the spectrum, until the overall area under the curve is the same as it was initially. Adding the CO_2 caused the planet to warm.

Take-home points

1. Gases absorb/emit IR light if they vibrate at the frequency of the light, and if its vibration has a dipole moment that affects the electric field. O_2 and N_2 are not greenhouse gases. All molecules of three or more atoms are IR active.
2. A greenhouse gas has a stronger impact on the radiative balance of the Earth if it interacts with light in the middle of the earthlight spectrum.
3. Band saturation: a greenhouse gas at relatively high concentration like CO_2 will be less effective, molecule per molecule, than a dilute gas like methane.

Projects

Answer these questions using the online model at http://understandingtheforecast.org/Projects/infrared_spectrum.html. The model takes CO_2 concentration and other environmental variables as input, and calculates the outgoing IR light spectrum to space, similarly to Figs. 4.3, 4.5, and 4.8. The total energy flux from all IR light is listed as part of the model output, and was used to construct Fig. 4.6.

1. *Methane.* Methane has a current concentration of 1.7 ppm in the atmosphere, and it's doubling at a faster rate than is CO_2.

a. Is 10 additional ppm of methane in the atmosphere more or less important than 10 additional ppm of CO_2 in the atmosphere at current concentrations?

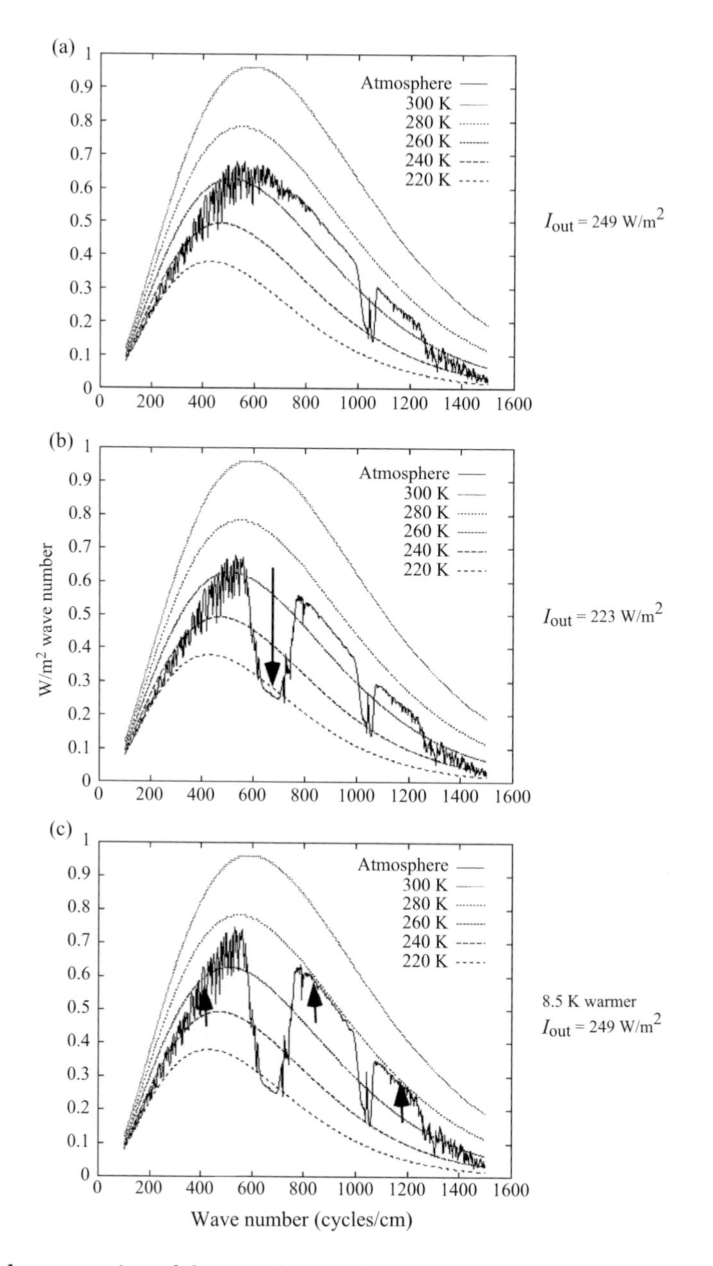

Fig. 4.8 **A demonstration of the greenhouse effect of CO_2. (a) We begin with no CO_2. Let's assume that the energy budget of the Earth was in balance at a ground temperature of 270 K. In (b) we add 1000 ppm CO_2, decreasing the outgoing energy flux. (c) The ground and the atmosphere above it respond by warming up to 8.5 K. The total outgoing energy flux is restored to its initial value. The total energy flux is proportional to the area under the curves. CO_2 takes a bite out of (a) to generate (b), but (c) bulks up everywhere to compensate.**

b. Where in the spectrum does methane absorb? What concentration would it take to begin to saturate the absorption in this band? (How do you identify saturation of a band on a spectrum plot?)
c. Would a doubling of methane have as great an impact on the heat balance as a doubling of CO_2?
d. What is the "equivalent CO_2" of doubling atmospheric methane? That is to say, how many ppm of CO_2 would lead to the same change in outgoing IR radiation energy flux as doubling methane? What is the ratio of ppm CO_2 change to ppm methane change?

2. *CO_2.*

a. Is the direct effect of increasing CO_2 on the energy output at the top of the atmosphere larger in high latitudes or in the tropics?
b. Set pCO_2 to an absurdly high value of 10,000 ppm. You will see an upward spike in the CO_2 absorption band. What temperature is this light coming from? Where in the atmosphere do you think this comes from?

3. *Earth temperature.* Our theory of climate presumes that an increase in the temperature at ground level will lead to an increase in the outgoing IR energy flux at the top of the atmosphere.

a. How much extra outgoing IR would you get by raising the temperature of the ground by 1°C? What effect does the ground temperature have on the shape of the outgoing IR spectrum and why?
b. More water can evaporate into warm air than cool air. By setting the model to hold the water vapor at constant relative humidity rather than constant vapor pressure (the default), calculate again the change in outgoing IR energy flux that accompanies a 1°C temperature increase. Is it higher or lower? Does this make the Earth more sensitive to CO_2 increases or less sensitive?
c. Now see this effect in another way. Starting from a base case, record the total outgoing IR flux. Now increase pCO_2 by some significant amount, say 30 ppm. The IR flux goes down. Now, using the constant vapor pressure of water option, increase the temperature offset until you get the original IR flux back again. What is the change in T required? Now repeat the calculation but at constant relative humidity. Does the increase in CO_2 drive a bigger or smaller temperature change? This is the water vapor feedback.

Further reading

The Discovery of Global Warming (2003) by Spencer Weart. This is a historical account of the science and the scientists who discovered global warming including my favorite, Svante Arrehnius, who used the IR spectrum of moonlight, in 1896, to predict that doubling CO_2 would raise global temperature by 3–6°C (whereas the modern prediction is 2–5°C). There is a good discussion of piecing together the band saturation effect in this book.

IPCC Scientific Assessment 2001 from Cambridge University Press or downloadable from http://www.grida.no/climate/ipcc_tar/. See Chapter 6 "Radiative Forcing of Climate Change."

5
The temperature structure of the atmosphere

> The layer model assumes that heat flows between the Earth's surface and the atmosphere by radiation. In reality, the temperature of the atmosphere is coupled to the temperature of the ground by convection. Warm air rises and carries heat. As it expands, it cools, leading to a decrease in temperature with altitude. If the atmosphere were incompressible, like water, there would be very little change in temperature with altitude, such as occurs in a pan of water heating up on a stove. If this were the case, then there would be no greenhouse effect because the outgoing IR intensity would be the same regardless of whether the light on average comes from the ground or from high in the atmosphere.

Meet the atmosphere

If you have ever climbed a mountain, you know that the temperature decreases as you go up (Fig. 5.1). This part of the atmosphere is called the ***troposphere***, and it contains 90% of the air in the atmosphere and all of its weather. If you were Superman and

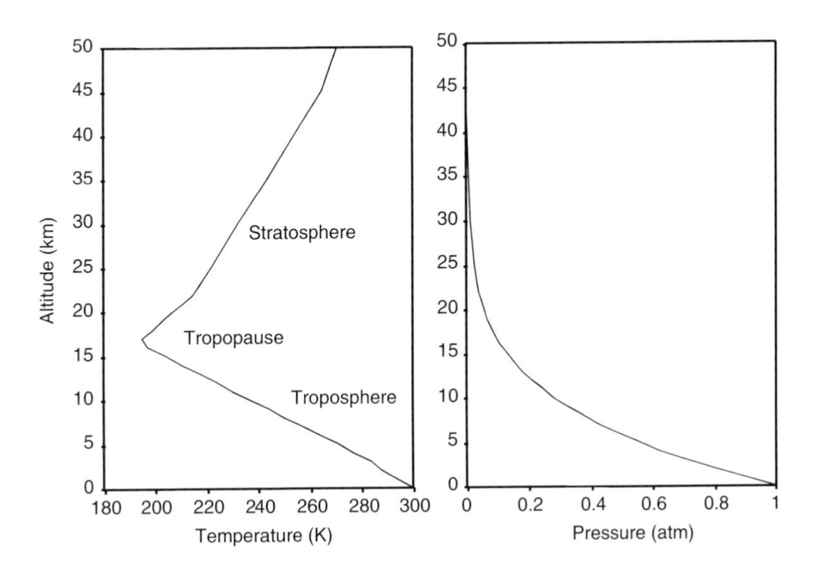

Fig. 5.1 **Typical temperatures and pressures of the atmosphere as a function of altitude in the tropics.**

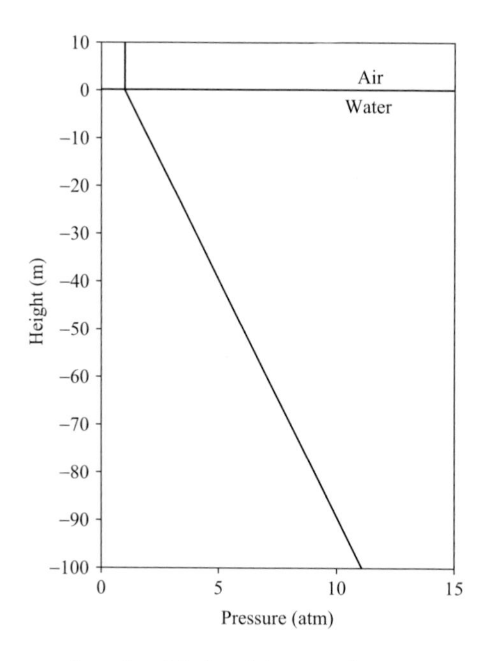

Fig. 5.2 **Pressure increases linearly with depth in water because water is incompressible.**

could climb high enough, the temperature would reach its coldest point at an altitude called the ***tropopause***, and would begin to rise above that in the ***stratosphere***. Above the stratosphere are wispier layers called the mesosphere and the exosphere which affect the way that ham radio signals propagate around the world, but don't affect our climate story very much.

Explaining the temperature structure of the atmosphere requires weaving together a number of what may seem to be disparate threads. Stick with me and everything will all tie together in the end.

Pressure as a function of altitude

Gases and liquids exert pressure on the surfaces of solids that are immersed in them, simply the force of the atoms bouncing off of the solid surface. The pressure gets lower as you climb higher in the atmosphere. As we ascend, we decrease the amount of fluid that is above us, decreasing the pressure that we feel.

Scuba divers know that diving 10 m deep increases the pressure by about 1 atm. Each 10 m of depth is the same 1 atm pressure increase: descending from 30 to 40 m would increase the pressure by the same 1 atm as descending from 0 to 10 m. We say that pressure is ***linear*** with depth (Fig. 5.2). The pressure can be calculated as

$$P = 1\,\text{atm} + \frac{-1\,\text{atm}}{10\,\text{m}} \cdot z[\text{m}]$$

where we are using variable z to denote the vertical position, as before, with positive numbers upward, as before. So a depth is a negative height. At the water surface, $z = 0$, and we have 1 atm pressure from the weight of the atmosphere. The increase in pressure with depth, from the weight of the water, is linear with depth below the surface because the factor -1 atm/10 m of height is constant.

The pressure in the atmosphere is ***nonlinear*** with altitude, in that a climb of 1 m at sea level changes the pressure much more than 1 m of climb up at the tropopause. The equation to describe pressure as a function of height in the atmosphere is based on the ***exponential function***, which is a number called e raised to a power. The value of e is approximately 2.71828... The exponential function was invented by bankers to calculate compound interest for bank accounts. The interest that a bank account pays is proportional to the amount of money in the account. The mathematical formula for the amount of money in the bank at any given time, if the payout from interest is continually deposited into the account to start earning interest itself, is the exponential function

$$\text{Balance}(t) = \text{Balance}(\text{initial}) \cdot e^{k \cdot t}$$

where Balance(t) is the bank balance at any time t in years, and k is an interest rate, like 10% per year or 0.1 year^{-1}.

The exponential function comes up time and again in the natural sciences. Population growth and radioactive decay are two examples. In each case, the rate of change of the variable depends linearly on the value of the variable itself. Population growth is driven by the number of babies born, which depends on the number of potential parents to beget them. The rate of radioactive decay that you would measure with a Geiger counter depends on the number of radioactive atoms present. The growth rate of your bank account depends on its size.

The multiplier in the exponent describes the relative rate of change of the variable, in this case, the interest rate. If the exponent is positive, as for population, the growth accelerates as it progresses (Fig. 5.3). One could reasonably call this type of growth an explosion. For decay, the exponent is negative, and the quantity of radioactive atoms gets ever closer to zero with time, but mathematically never gets there.

The atmospheric pressure varies as a function of altitude according to an exponential decay type of equation in height

$$P(z) = 1\ \text{atm} \cdot e^{-z[\text{km}]/8\ \text{km}}$$

The height z is zero at sea level, leaving us with e^0 which equals 1, so the pressure at sea level is 1 atm. At an altitude of 8 km, pressure is lower than at sea level by a factor of e^{-1} or 1/e, about 37%. We call that altitude the ***e-folding height***. Most of the mass of the atmosphere is contained in the e-folding height. In fact, if the entire atmosphere were to be at 1 atm pressure, instead of smeared out in a decaying exponential function, it would fit into one e-folding height exactly. If we were tracking the decay of a radioactive chemical with time, the scaling factor in the exponential would be called an ***e-folding time***. This quantity is similar to a half-life for radioactive decay but rather than decaying to half the initial quantity, we're waiting until we're 37%

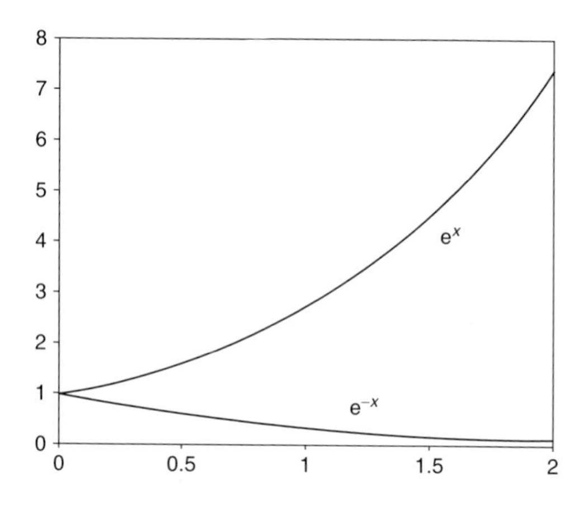

Fig. 5.3 **The exponential functions e^x and e^{-x}. For e^x, the growth rate of the function is proportional to its value. Examples of this type of behavior include interest on a bank account and population growth. For e^{-x}, the value of the function decays proportionally to its value. Radioactive decay does this, as does pressure as a function of altitude in the atmosphere.**

lower than the original. It takes 44% longer to decay to 1/e than it does to 1/2 of the original number of atoms.

From the appearance of the exponential function in the equation for pressure, you could probably guess that the rate of change of pressure with altitude must depend on the pressure itself in some way. This would be astute. The rate of change of pressure depends on pressure because at high pressure, gas is compressed, and so a climb of 1 m through gas at high pressure would rise above more molecules of gas than would a climb through a meter of gas at low pressure. Imagine a wall made of compressible bricks (Fig. 5.4). A row of bricks is thinner at the bottom of the wall because they are compressed. Batman climbing up the wall would pass more bricks per step at the bottom than at the top. For incompressible bricks (the normal kind), the rows are all of the same height and the mass of the wall above you is a linear function of height.

Adiabatic expansion

Here begins the second apparently unrelated thread of our story. If we compress a gas, its temperature goes up. This occurs even if we don't allow any heat to enter the gas or leave it, say if we had gas inside an insulated piston that we compress or expand. The condition that we are describing, a closed system with no heat coming in or out, is called ***adiabatic***. If gas is compressed adiabatically, it warms up. If you ever let the air out of a bicycle tire by holding the little valve open with your thumbnail, you may have noticed how cold your thumbnail got. The gas expanded as it flowed from the high-pressure tire into the lower-pressure atmosphere.

Figure 5.5 shows the temperature change that a parcel of dry surface air would experience if it were carried aloft adiabatically (an ***adiabatic trajectory*** or ***adiabat***).

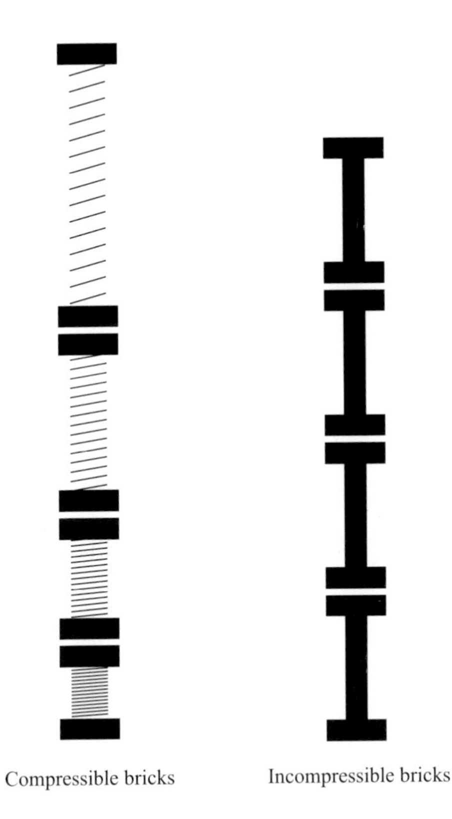

Fig. 5.4 **A demonstration of why pressure changes rapidly at the bottom of the atmosphere and slowly at the top, because air in the atmosphere is compressible (left). For water, which is incompressible, pressure increases linearly with depth.**

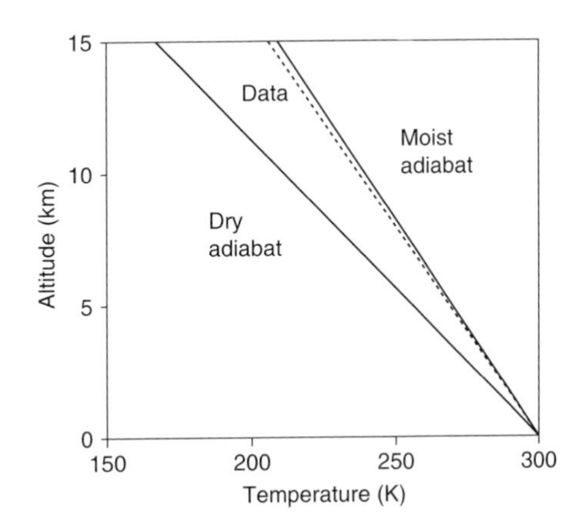

Fig. 5.5 **The temperature decrease with altitude in the atmosphere is called the lapse rate. If the atmosphere contained no water vapor, the lapse rate would follow the dry adiabat. As is, most of the atmosphere is closer to the moist adiabat, especially in the tropics where convection is most important.**

We'll worry about water vapor in a moment, but for now we are looking at the line labeled dry adiabat (foreshadowing).

Why should the temperature go up in a gas as you compress it? It takes work to compress a gas. You have to squeeze our hypothetical insulated piston in order to compress it. Your muscles push against molecules bouncing against the walls of the cylinder. The work you put in is transformed into bouncing-around energy of the molecules of gas, its temperature. The situation is a little harder to envision for expansion because we tend to ignore the atmosphere around us and think of an empty room as empty space, but when the piston expands, it has to push back the atmosphere, in other words it must do work. The energy to do that work comes from the thermal energy of the molecules of the gas, so the gas cools as it expands.

Water vapor and latent heat

Here comes seemingly unrelated thread number three. Water molecules can exist together in any of three ***phases***: gas, liquid, or solid. A transformation from liquid or solid to gas requires an input of a considerable amount of energy. One could write a chemical reaction for water as

$$\text{Vapor} = \text{Liquid} + \text{Heat}$$

If you have ever burned your skin with steam from a teapot, this reaction will have meaning for you. Steam from a teapot is probably at the boiling temperature of water, 373 K. This is a temperature you might set for your oven to warm bread; you can easily blow the air from the oven on your hand when you reach in for the bread without burning yourself. Steam burns, not because it is so hot, but because it deposits its heat when it condenses on your skin. The heat that the steam is carrying is called ***latent heat***. You charge up an air parcel with latent heat when you evaporate water into it, and you get the heat back again when the water condenses. A thermometer doesn't measure the latent heat content of an air parcel unless the heat is released by condensation.

If you set a tray of water or ice in a closed box with some extra space around it, some of the water molecules are going to evaporate, becoming a gas. Here's an interesting tidbit; it makes almost no difference whether there are other gases in that space or just vacuum. The same number of water molecules will make the jump into the gas phase regardless; that is to say, the pressure arising from water vapor doesn't care how much oxygen or nitrogen is there. The pressure from water vapor depends primarily on the temperature. If the temperature is high, many molecules will have the energy required to jump into the vapor phase, and the vapor pressure of water will be high. If it is cold, the vapor pressure will be lower (Fig. 5.6).

At any given temperature, the water vapor pressure will drift toward some ***equilibrium*** pressure. If the amount of water vapor is lower than equilibrium, that is to say ***undersaturated***, water will tend to evaporate. If water vapor is ***supersaturated***, meaning the pressure is higher than equilibrium, then water will tend to condense, perhaps into raindrops or snowflakes. Equilibrium is the state of lowest energy; a dead

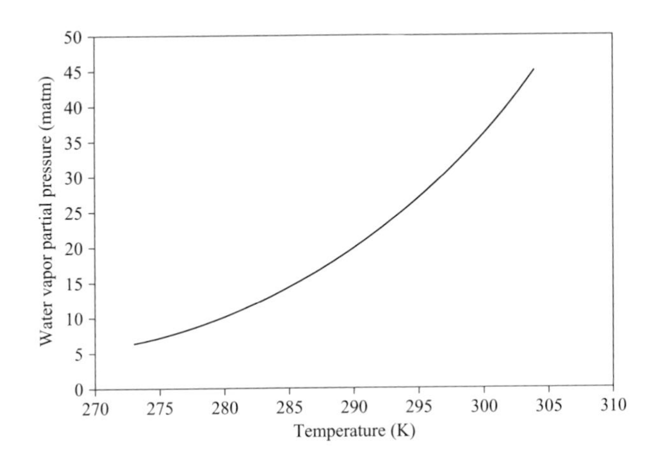

Fig. 5.6 **Water has a greater tendency to evaporate at higher temperature than at a cooler temperature. The graph shows what a chemist would call the equilibrium vapor pressure of water, and a meteorologist would call 100% relative humidity. Air in the real world can have less or sometimes slightly more water vapor than this, but if the water vapor concentration gets too much higher than this curve, the water will condense into water droplets or snow.**

battery is in equilibrium. However, a chemical system can be out of equilibrium for a long time if the reaction rates are very slow. In the clean remote marine atmosphere, forming droplets can be very slow, and water vapor 25% supersatured has been observed. The ***relative humidity*** is the water vapor pressure divided by the saturation value; from the last example, 25% supersaturated would be 125% relative humidity.

Convection

Now we are ready to weave our seeming disparate threads of story together into a picture of what controls the temperature as a function of altitude in the atmosphere. The pieces are assembled into a process called ***convection***. Convection takes its place among conduction and radiation, which we have already discussed, as a means of carrying heat in the environment. Convection occurs when a fluid medium is heated from the bottom or cooled from the top, and the heavy water atop light water causes the fluid to circulate. Imagine a lava lamp, in which the fluid is heated from below by a light bulb (a ridiculous energy-wasting incandescent bulb; you could never make a lava lamp work with a nice, efficient, compact fluorescent bulb because fluorescent bulbs do not generate so much waste heat). The fluid at the bottom becomes warmer than the fluid overlying it. Warmer molecules bounce around more energetically, pushing all of the molecules somewhat farther apart from each other. For this reason the fluid ***expands*** as its temperature increases. As it expands, its density (mass per volume) decreases. If we stack some fluids of different density together in a column, the stable configuration is to have the densest ones on the bottom. Think of oil and water; the oil always floats on the water because it is less dense. So our expanding parcel of fluid at

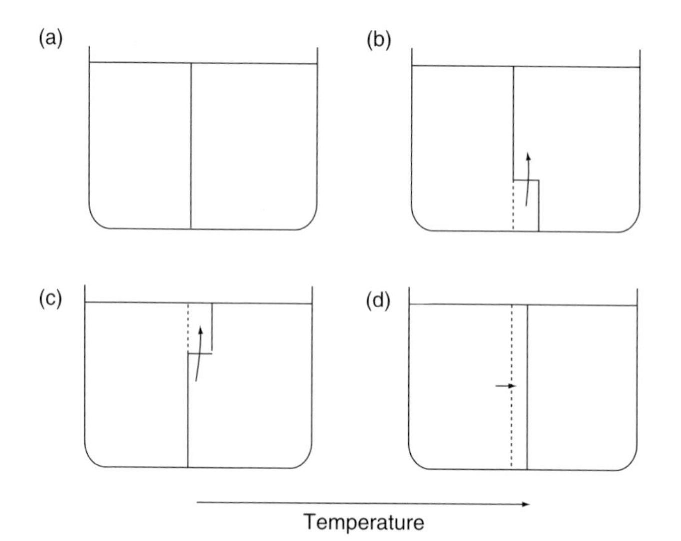

Fig. 5.7 **The effect of convection on the temperature in a pan of water on a stove. In (a: static stability) the water in the pan is well mixed, the same temperature throughout. After the burner is turned on, the water at the bottom of the pan warms, which makes it less dense than the water above it (b: convectively unstable). Either this water can rise to the top without mixing (c: stratified) or it can mix with the water above it, raising the temperature of the entire pan of water (d: new static stability). The atmosphere tends to mix when it convects, as in (d).**

the bottom of the lamp begins to rise. It floats to the top of the lamp until it cools and sinks back to the bottom again.

Forget the lava lamp now and think of a pan of water on a stove, which is a simpler case because there is only one fluid instead of two. Figure 5.7 plots the temperature of the water as a function of the height in the pan, which we call a temperature ***profile.*** If we thoroughly mix the water, the temperature will be the same throughout the pan of water. This sounds obvious but we'll find in a moment that when we mix a column of air, the temperature is not uniform throughout. A well-mixed, uniform temperature water column is called ***statically stable*** because if the fluid is left alone, it won't feel any need to circulate. Any parcel of water has the same density as its neighbors.

Next we'll turn on the burner, warming and expanding the water at the bottom of the pan. Buoyant water at the bottom tends to rise: this situation is called ***convectively unstable.*** The buoyant water from the bottom could rise to the top like the lava in the lava lamp, in which case we would end up with denser water underlying lighter water, which we call ***stratified.*** Pubs in Ireland serve a concoction called a black and tan, with warm dark Guinness Porter overlying cold Bass Ale. The two types of beer remain unmixed because they are stratified by temperature. Alternatively (back to the pan analogy, alas) the rising warm water could mix with the rest of the water, as generally occurs to a large extent in the atmosphere and ocean. In this case we end up with a second statically stable profile at a higher temperature than the first.

Convection in a compressible gas is analogous to convection in an incompressible fluid like our pan of water (Fig. 5.8). We begin from the temperature profile of static

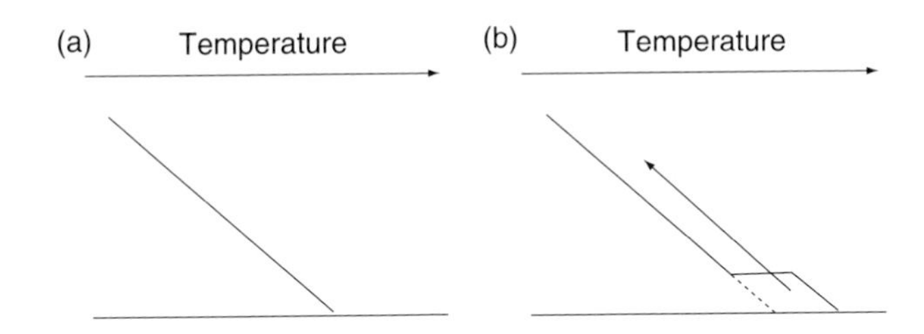

Fig. 5.8 **Convection in a compressible fluid like air. (a) is the statically stable configuration, analogous to (a) in Fig. 5.7. The temperature decreases with altitude because the gas expands, and therefore cools. This temperature profile is what you would see in a well-mixed column of air. (b) is the effect of heating from below, analogous to (b) in Fig. 5.7.**

stability. As for the water column, we can construct a statically stable temperature profile by mixing the gas thoroughly. No gas parcel will be more or less dense than its neighbors because it is all the same stuff. The pressure decreases as you ascend the gas column, and so a gas parcel raised from the bottom of the column expands, and its temperature drops. After it does so, our parcel finds that it is still exactly the same temperature as the gas it finds itself surrounded by. The parcel was neutrally buoyant at the bottom of the column, and it is neutrally buoyant aloft.

Convection is driven by heating at the bottom of the column, such as by sunlight hitting the ground. The warmed air parcel from the ground finds itself warmer than the air immediately above it, even when it has expanded to the pressure of the air above it. The rising parcel follows its own adiabat, which is higher than the temperature profile of the gas column, and so the rising parcel has the ability to rise to the top of the gas column if it does not mix with other gas on the way up. If the gas does mix, the temperature profile of the entire column will rise to a new adiabat, all completely analogous to the incompressible case of the pan of water.

Moist convection

Convection in a column of a compressible gas is more complicated than it was in our nice simple pan of incompressible water, but I hope it helped to think about the incompressible water case first. To get to convection in the real atmosphere, though, we need to add one more ingredient, and that is latent heat from water vapor. Imagine a source of water vapor at the Earth's surface, so that the air above the ground has a relative humidity of 100%. If we raise this air parcel aloft, it will decompress and cool. The saturation water vapor pressure decreases with temperature (Fig. 5.6), and so the air parcel becomes supersaturated, carrying more water than it wants for equilibrium. The excess water condenses into cloud droplets or perhaps rain. As the water condenses, it releases its latent heat, just as the steam from the teapot burns your skin by releasing latent heat. The effect of the latent heat is to warm the air parcel up a bit in the face of its decompression cooling (Fig. 5.9). We distinguish between the dry and the wet cases by referring to a ***dry adiabat*** or a ***moist adiabat***. One more definition: the decrease in

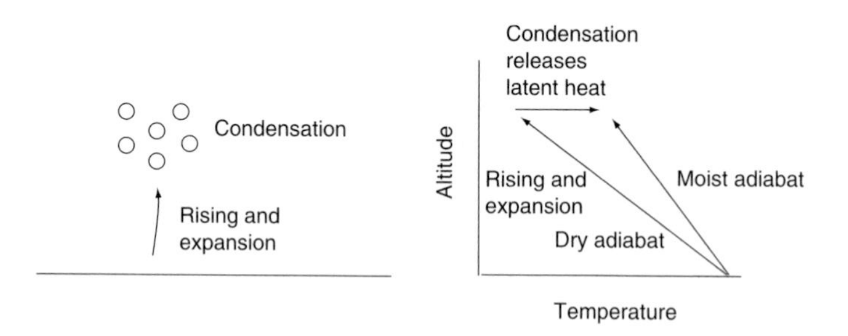

Fig. 5.9 **The effect of water on the temperature profile in the atmosphere. Air cools as it rises, as in Fig. 5.8. The cooling drives water vapor to condense into droplets, releasing its latent heat to the air. The moist adiabat therefore cools less with altitude than does the dry adiabat.**

temperature with altitude in the atmosphere is called the ***lapse rate***. The lapse rate of a dry convective atmosphere would be about 10 K/km of altitude, whereas the lapse rate of a wet atmosphere is only about 6 K/km.

Convection in the layer model

The layer model that we constructed in Chapter 3 did not have convection. Think about a layer model with multiple atmospheric layers, such as one constructed in Project 2 in Chapter 3. The temperatures of the layers decrease with altitude, just like the real atmosphere does, but the only way that heat is carried between the layers in the layer model is by blackbody radiation. One could construct a model with a continuous atmosphere, with temperature varying smoothly with altitude like the real atmosphere, but where radiation is the only way of moving heat. The temperature profile you would get from a model like this is known as ***radiative equilibrium***. Radiative equilibrium controls the temperature structure inside some stars and would control the temperature profile in the Earth's atmosphere, except that convection kicks in first. A radiative equilibrium lapse rate in our atmosphere would be about 16 K/km of altitude, so steep that it would be convectively unstable.

If we were to try to add convection to the layer model, we would have to add another set of heat flow arrows to the model, representing the heat carried by air as it ascends, and by water vapor as it condenses releasing latent heat. We are not going to create such a model, but Fig. 5.10 gives an impression of how it might look. Convection might insist that the temperature of layers aloft must follow a moist adiabat, roughly 6 K/km of altitude. The state of energy balance when both radiation and convection are taking place is called ***radiative–convective equilibrium***.

Lapse rate and the greenhouse effect

The steeper the lapse rate, the stronger the greenhouse effect. If the atmosphere were incompressible like water, and convection maintained a uniform temperature with

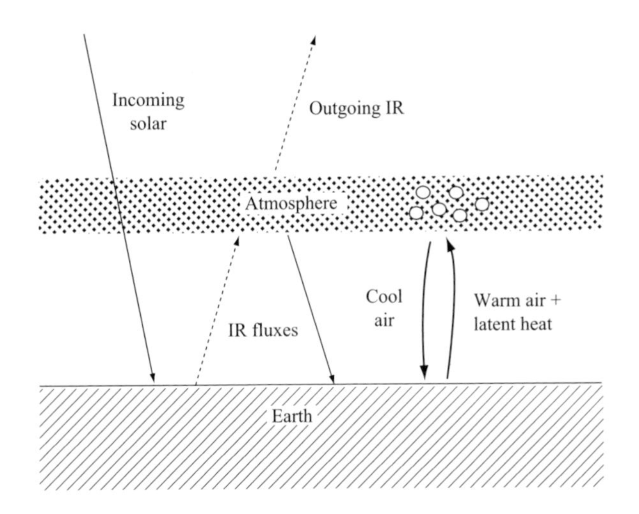

Fig. 5.10 **The layer model from Chapter 3 as it might look if we were to include convection. Convection carries heat vertically in the atmosphere, supplementing the heat carried by radiation. If we were to add convection to the layer model, it would require a few new arrows.**

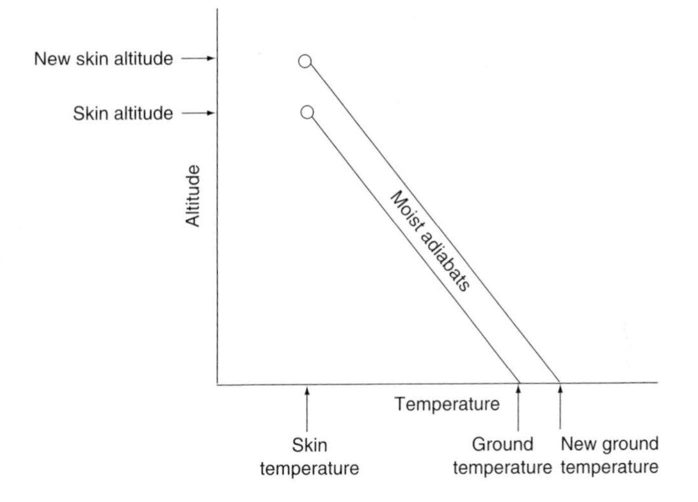

Fig. 5.11 **A demonstration of the effect of the lapse rate on the strength of the greenhouse effect. If we increase the greenhouse gas concentration of the atmosphere, the IR radiation to space will originate from a higher altitude (labeled skin altitude), but the skin temperature at the skin altitude will be the same as before. The increase in skin altitude increases the ground temperature. If the temperature of the atmosphere were the same at all altitudes, then raising the skin altitude would have no impact on ground temperature.**

altitude, as it did in our pan of water on the stove, there would be no greenhouse effect. To understand why this is so, imagine increasing the concentration of CO_2 in an atmosphere. This has the effect of raising the altitude in the atmosphere where light on average escapes to space (Fig. 5.11). You remember from Chapter 4 that reality is a bit complicated in this regard; some IR light goes directly to space from the ground, in the frequency range of the atmospheric window, while at other frequencies, like in the CO_2 band, light appears to originate from the coldest part of the atmosphere, at the tropopause. But bunching all of that light together in our minds, we can imagine that, on average, light comes from a higher altitude as the CO_2 concentration goes up. Let's call that altitude the ***skin altitude***.

Now think back to Chapter 3 and the layer model. Remember that the outermost part of the atmosphere, the part that radiates directly to space, always had the same temperature in all of those different model configurations. We called that the skin temperature, and it was always 253 K for the albedo and sunlight intensity of the Earth, whatever the model configuration. In solving the layer model for the temperatures of all the atmospheric layers and the ground, it was convenient to start from the outer skin layer and work down toward the ground. Let's take that approach again, assuming that the temperature change with altitude (the lapse rate) has been decided by the physics of convection. If the skin altitude were 5 km high and the lapse rate were 6 K/km, then the temperature of the ground would be

$$T_{\text{ground}} = T_{\text{skin}} + \frac{6\,\text{K}}{\text{km}} \cdot 5\,\text{km}$$

We can visualize this as a line drawn downward from the skin altitude, following the slope of a moist adiabat until it intersects the ground (Fig. 5.11). If we increase the CO_2 content of the atmosphere, we raise the skin altitude, and the same moist adiabat slope intersects the ground at a higher temperature. Algebraically, the change in temperature from raising the skin altitude z_{skin} can be calculated as

$$\Delta T = \Delta z_{\text{skin}}[\text{km}] \cdot \frac{6\,\text{K}}{\text{km}}$$

Here's the point. If the lapse rate were different than 6 K/km, then a change in CO_2, driving a change in the skin altitude, would have a different effect on the temperature of the ground. If the atmosphere were incompressible, for example, and convection insisted that the temperature should be uniform with altitude, then it would make no difference how high the skin altitude was:

$$\Delta T = \Delta z_{\text{skin}}[\text{km}] \cdot \frac{0\,\text{K}}{\text{km}} = 0$$

The temperature of the ground would be the same as the skin temperature no matter what.

The lapse rate in the atmosphere (how quickly temperature decreases with altitude) is determined primarily by convection and the "hidden" heat carried aloft by water vapor. The lapse rate determines the sensitivity of the temperature of the ground to changes in the IR opacity of the atmosphere. If we want to forecast the effect of the rising atmospheric CO_2 concentration on the temperature of the ground, we will have to get the lapse rate right, and any changes in the lapse rate that result from future climate change.

Take-home points

1. Pressure decreases with altitude.
2. Temperature decreases as a gas expands.
3. Moisture in the rising gas releases its latent heat as it condenses.
4. The lapse rate is controlled by the moist adiabat.
5. The strength of the greenhouse effect depends on the lapse rate.

Projects

1. *Lapse Rate.* Use the online full-spectrum radiation model at http://understanding-theforecast.org/Projects/full_spectrum.html. Adjust the lapse rate in the model and document its impact on the equilibrium temperature of the ground.

2. *Skin Altitude.* Answer this question using the online IR radiation model.

a. Run the model in some configuration without clouds and with present-day pCO_2. Compute σT^4 using the ground temperature to estimate the heat flux that you would get if there were no atmosphere. The value of σ is $5.67 \cdot 10^{-8}$ W/m^2 K^4. Is the model heat flux at the top of the atmosphere higher or lower than the heat flux you calculated at the ground?

b. Now calculate the "apparent" temperature at the top of the atmosphere by taking the heat flux from the model and computing a temperature from it using σT^4. What is that temperature, and how does it compare with the temperatures at the ground and at the tropopause? Assuming a lapse rate of 6 K/km, and using the ground temperature from the model, what altitude would this be?

c. Double CO_2 and repeat the calculation. How much higher is the skin altitude with doubled CO_2?

d. Put CO_2 back at today's value, and add cirrus clouds. Repeat the calculation again. Does the cloud or the CO_2 have the greatest effect on the "skin altitude"?

Further reading

Frederick, John E., *Principles of Atmospheric Science*, Jones and Bartlett Publishing, Inc., Sudbury, MA, 2007.

6
Heat, winds, and currents

Unlike the layer model, the ground temperature in the real world is not the same everywhere. The seasonal cycle drives surface temperatures to values above and below the average value. Also, unlike the layer model, energy budgets such as in the layer model do not balance locally even if you average out the time-varying part. There is a net heat input in low latitudes and transport to high latitudes. This process of heat transport is complex, driven by the turbulent flow of air and water on the rotating Earth, requiring complex computer climate models.

Averaging

The layer model assumes that heat fluxes in and out of the Earth must be in balance exactly. What the layer model is after is an average temperature over the entire globe and over time. This is a reasonable assumption to make if we are willing to wait long enough for our answer to be right. On long enough timescales there is simply nowhere else for the heat energy to go; what comes in must go out. As we look in more detail, however, there are all kinds of wild imbalances. It takes heat every spring to warm lakes and ocean water and to melt the snow. There is also an imbalance between the heat fluxes on the long time-average, because of heat transport from equatorial regions and the higher latitudes to the north and south.

So is an eternal, unchanging, averaging model a reasonable one for a world that is bouncing around like Jell-O? Can you construct the average of the whole system by using averages of the pieces of the system? Or will averaging change the answer? The term for this possibility is ***biasing***. Biasing issues come up a lot in the natural sciences. In principle, there could be a problem with averaging IR energy fluxes because they are a nonlinear function of temperature. One way to see the nonlinearity is to look at the equation and see that the light flux is proportional to temperature to the fourth power, and not to the first power, which would be linear

$$I\left[\frac{\mathrm{W}}{\mathrm{m}^2}\right] = \varepsilon[\text{unitless}]\,\sigma\left[\frac{\mathrm{W}}{\mathrm{m}^2\mathrm{K}^4}\right]T[\mathrm{K}]^4$$

Another way would be to notice that a plot of the function is not a straight line (Fig. 6.1). Let's say we wanted to estimate the average energy flux of a planet that had two sides, one at 300 K (rather like Earth) and the other at 100 K (much colder than anywhere on Earth). The outgoing energy flux from the cold side would be about 6 W/m^2, and from the warm side 459 W/m^2. The average energy flux would be 232 W/m^2. Let's now

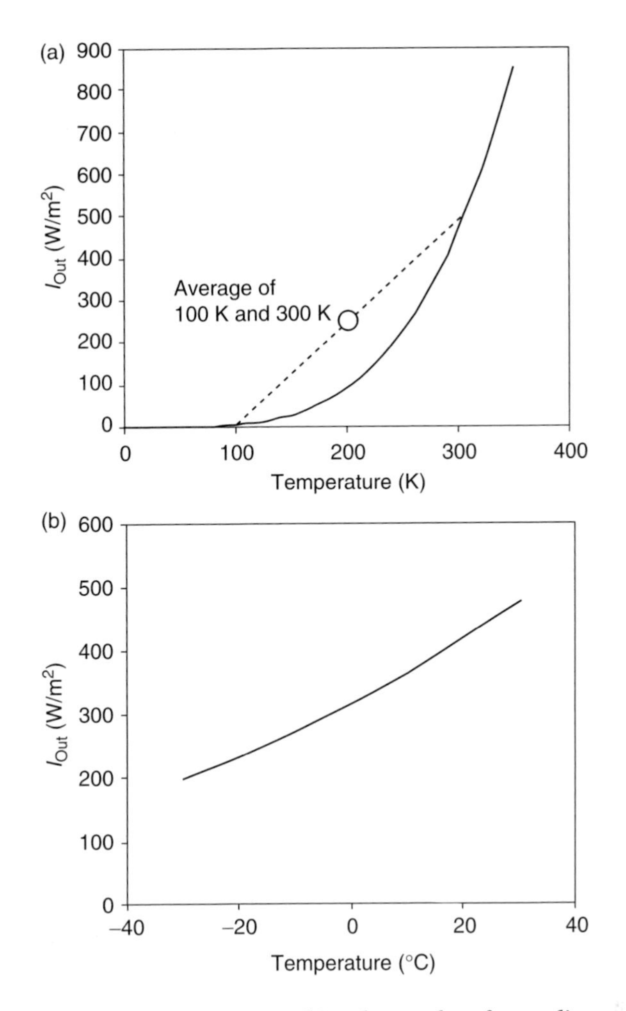

Fig. 6.1 **An example of how averaging can bias the results of a nonlinear system such as blackbody radiation energy flux σT^4. In (a) we average across a huge temperature range, and the flux we get from combining areas of 100 and 300 K is very different from what we would expect if we averaged the temperature first to 200 K and then computed the answer. In (b) we see that over the temperature range of normal Earth conditions, the blackbody radiation energy flux is closer to linear, so averaging would be less of a problem, than in (a).**

average the temperatures, run the average T through σT^4, and try to get the same answer. The average temperature is 200 K, and the predicted energy flux is 91 W/m^2. We'd be off by more than a factor of two. You can see the effect graphically as the difference between the straight line and the curve in Fig. 6.1a.

For the terrestrially normal range of temperatures (Fig. 6.1), we are zooming in on the nonlinear function enough that it looks much straighter than when we considered an absurdly wide temperature range. Still, it could be an important effect, and we certainly wouldn't want to neglect it in the global warming forecast. It doesn't appear that the biasing is so bad as to undermine the principles demonstrated by the layer model, however, because in this case the function turned out to be fairly linear.

Biasing could arise when calculating a time average from a time series of measurements, as if air temperatures were measured more often during daytime than at night, for example. The problem can be corrected for, as long as it is recognized. There are many other ***nonlinearities*** in the climate system, pairs of variables that are related to each other in nonlinear ways. For example, many effects of the wind, such as exchanging heat with the ocean, depend nonlinearly on wind speed.

Weather versus climate

If you don't like the springtime weather in Chicago, just wait a few days. It's a gray 5°C out there now, but the forecast for the weekend puts it up to 15°C, which is a little more like spring. The 10-day forecast says showers the weekend after that, but no one believes the end of a 10-day forecast anyway. They're better than they used to be, but 10 days is still something of a crap shoot. And here I am sitting down to write about forecasting the climate 100 years from now. I suppose some would feel that an explanation might be in order.

It is indeed tricky to forecast the ***weather*** too far in advance because weather is ***chaotic***. To a scientist, the word chaotic brings to mind an extreme sensitivity to initial conditions, so that small differences between two states tend to amplify, and the states diverge from each other. This behavior is called the ***butterfly effect***, the reference being to a puff of air from a butterfly's wing eventually resulting in a giant storm somewhere that wouldn't have happened if the butterfly had never existed. The butterfly effect was first observed in a weather simulation model. The model computer run stopped, and the researcher Edward Lorenz restarted it by typing in the values of model variables like temperatures and wind speeds, but with a small seemingly insignificant change of initial conditions. It didn't take long for the restarted simulation to diverge from the results of the initial simulation. The weather is forecast by constructing an initial condition from meteorological data and past model results, and running this initial condition forward into time using a model. The initial condition will never be perfect, and these imperfections tend to blow up, so that by about 10 days the prediction becomes meaningless. One way to cope with this situation is to run the model lots of times with tiny variations in initial conditions; an ***ensemble*** of model runs. Then you can see something of the range of what's possible in the forecast. It doesn't fix the problem exactly but it does result in a more reliable forecast.

The ***climate*** is defined as the time average of the weather. One could speak of a ***climatological*** January, which would be the average of many Januaries, let's say 10 Januaries. Forecasting climate is not as difficult as forecasting weather because it doesn't matter whether the rain you predict comes on a Tuesday or a Thursday. The butterfly can't foil our climate prediction with her storm, if the model gets the large-scale tendency for storminess right. If the model doesn't follow a storm trajectory this time, it will the next time; the butterflies average out. You would also like your climate model to get the frequency of the most extreme events, etc., what climatologists would call getting the ***statistics*** right. Even the decade and longer timescale "climate" output from models is sensitive to initial conditions, however, and so the ensemble technique is used

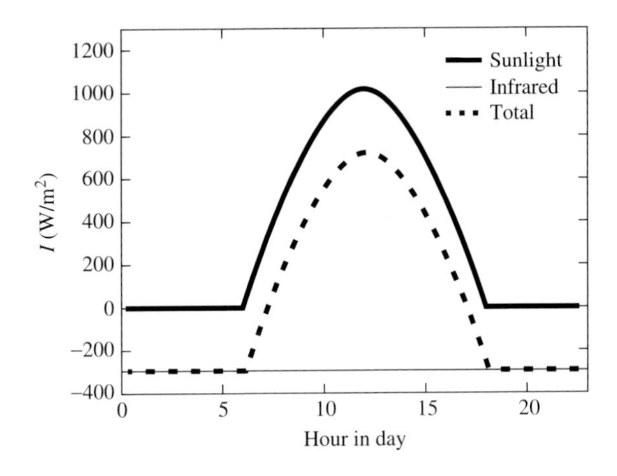

Fig. 6.2 **The surface of the Earth only receives incoming solar radiation during the daytime (heavy solid line), but it radiates IR light all the time (thin solid line). The energy budget for this location (dashed line) is only in balance when it is averaged over 24 h.**

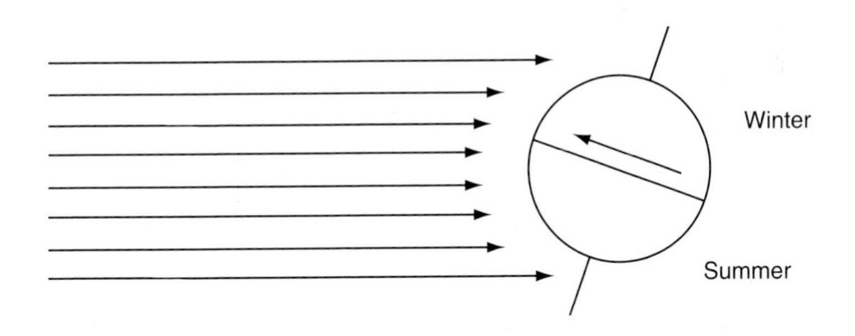

Fig. 6.3 **The Earth's tilt is responsible for the seasons. This is southern hemisphere summer, northern hemisphere winter.**

for climate forecasting, predictions to 2100, for example, just as they do for weather forecasting.

The heat-input forcing to the real world does not sit unchanged at the eternal average value, but rather bounces around pretty wildly. On the day–night cycle, for example, the night-time side of the Earth receives no solar input at all, while infrared (IR) energy loss to space continues around the clock (Fig. 6.2).

The heat input varies over the seasonal cycle as well. The ***seasons*** are caused by the tilt of the Earth's axis of rotation (Fig. 6.3). The issue is not the intensity of the incoming sunlight at the top of the atmosphere, I_{solar}, because the size of the Earth is negligible compared with the distance from the Sun. The Earth's orbit is elliptical, rather than circular, with the Earth closer to the Sun in southern hemisphere summer. However, if the distance from the Sun caused the seasons, the whole world would get cold and warm at the same time. The fact that the seasons are opposite each other in the two hemispheres proves that it is the tilt of the Earth, not the distance from the Sun, that causes the seasons.

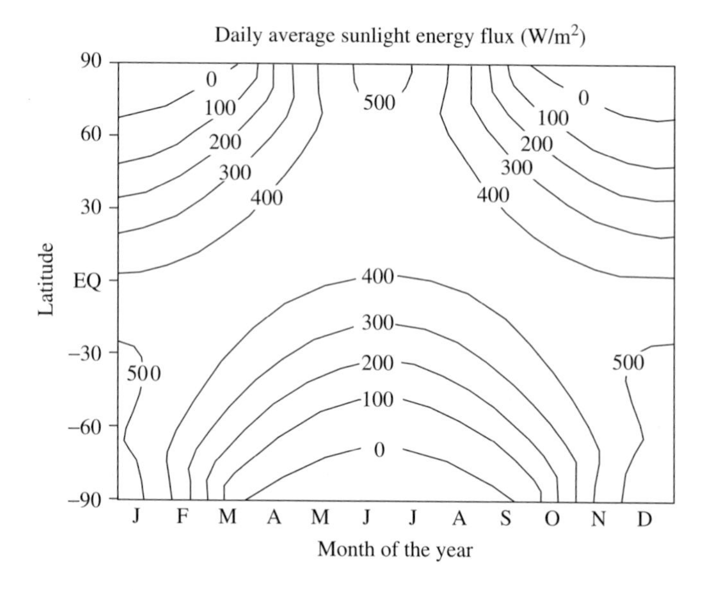

Fig. 6.4 **The Earth's tilt determines how much heat the surface receives from the Sun each day, as a function of latitude (vertical axis) and time of year (horizontal axis).**

The summertime hemisphere, the south in Fig. 6.3, gets more than its share of the solar energy, and the winter hemisphere gets less. Figure 6.4 shows a sort of map of the intensity of solar heating as a function of latitude and day of the year. The contours show us the intensity of sunlight for every square meter on the ground averaged over 24 h. Because the Earth completes one rotation in 24 h, any location on some line of latitude, say 42°N which goes through Chicago, will get the same solar influx as any other at that same latitude, like Barcelona or Vladivostok. A map of solar influx in regular latitude and longitude would look like a bunch of horizontal stripes. Therefore, we are showing two "dimensions" over which the solar influx does vary; latitude and day of the year.

The beginning of the year is southern hemisphere summer, and we see high heat fluxes in the south and low fluxes in north. Northern hemisphere summer is the middle of the year, centered around day 180. The pattern arises by two mechanisms. First, the intensity of sunlight per square meter of ground area is greater at noon in the summer than in the winter because the ground is at less of an angle to the Sun in the summer (recall Fig. 6.3). Second, days are longer in summer, and this increases the 24-h average energy influx.

It is interesting to note that the highest daily-average energy fluxes you find anywhere are at the north or south pole during north or south summer. The Sun never sets in midsummer at the poles, it just whirls around in a circle over the horizon. In winter the poles gets no sunlight for months on end. The poles do not turn into tropical garden spots in summer because it takes time for temperatures to respond to changes in heat forcing. This ***thermal inertia*** tends to damp out the temperature swings of the day–night cycle, the seasonal cycle, and any global warming temperature trends

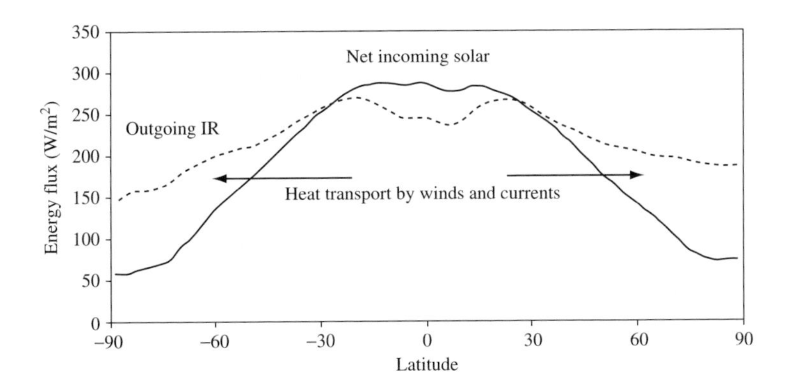

Fig. 6.5 **The energy budget between incoming solar and outgoing IR radiation does not balance locally because heat is transported on Earth by winds and currents. The equator receives more solar energy than it radiates as IR, while high latitudes lose more IR than they receive from sunlight.**

as well. The liquid ocean, it turns out, has much greater temperature stabilizing power than land or ice does because water mixes bringing a much greater mass of water into thermal exchange with the atmosphere than land can bring. Cooling of surface waters drives convection, in the same way that warming of surface air brings convection in the atmosphere. The seasonal cycle of temperatures in soils reach only a few feet down, while the temperatures farther down, in caves for example, reflect the yearly average, about 10°C in my part of the world. The seasonal cycle of temperature changes at the ocean reaches a 100 m down and more. For this reason, the seasonal cycle is much more intense in the middle of large continents than it is in "maritime" areas impacted by the temperature stabilizing effects of liquid water.

Even if we average out the seasonal cycle, the energy budget of a local region of the Earth may be far from being in balance because heat energy is redistributed around the Earth's surface by wind and water currents (Fig. 6.5). There is a net influx of heat in the tropics; sunlight brings in energy faster than it is radiated back to space. Heat is carried poleward by flow in the atmosphere and the ocean. In high latitudes, the Earth vents the excess tropical heat as excess radiative heat loss to space over direct solar heating.

So we have to simulate the weather

The upshot of this chapter is that in order to forecast global warming, we will have to simulate the time and space variations and imbalances in the energy budget, and the way that the Earth's climate responds to this forcing by storing or transporting heat around. The layer model will not do. Unfortunately, the physics which governs the flow of air and water is complex and difficult to simulate by its very nature. The difficulty is that fluid flow such as in the atmosphere and the ocean takes place on a wide range of size scales. Let's say we want to simulate the flow of the Gulf Stream in the ocean. The Gulf Stream is like a giant river in the ocean. Pieces of the flow spin off into the waters on

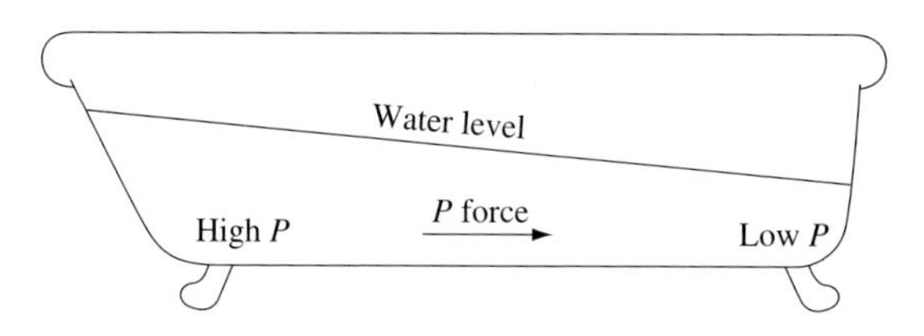

Fig. 6.6 **A thought experiment in which the bottom of a bathtub is level, that is to say, is on a geopotential surface, but the water level at the surface is sloped. In this situation there would be higher pressure on the geopotential surface at the level bottom of the bathtub on the deeper side than on the shallower side. This results in a pressure force that tends to push water from deeper to shallower.**

either side. Satellite images of ocean plankton distribution show rings, meanders, wisps, and eddies. The energy of the flow is being converted from the large-scale Gulf Stream into smaller-scale wisps and eddies. In the atmosphere, the large-scale flows break up into storms and weather. Eventually when the eddies get small enough, their energy is absorbed by friction of the flowing fluid. Some fluids like molasses have more friction than others like water; they are said to have higher ***viscosity***. This process of large-scale flow breaking up into smaller and smaller eddies until they get so small that they are absorbed by friction in the fluid is called the ***turbulent cascade***. The physicist Lewis Richardson summed it up in verse:

> Big whorls have little whorls,
> Which feed on their velocity,
> And little whorls have lesser whorls,
> And so on to viscosity.

At the heart of it, fluid flow is governed by Newton's laws of motion. Because fluid has mass it has ***inertia***, which tends to keep it moving if it's moving, or keep it stationary if it is already stationary. To change the speed or direction of the flow motion requires a force, such as gravity or pressure. One example of pressure driving fluid flow would be to fill up a bathtub with water and then by magic pull up on the water surface on one side, making a slope (Fig. 6.6), and then let it go. The water will flow to flatten out the surface. The way to see the pressure force is to think about the pressure within the fluid relative to the gravity field of the Earth. When we say "downhill" we mean downhill relative to a surface that is flat, which we call a ***geopotential surface***. The surface of a pool table had better be a good geopotential surface, or the balls won't roll straight. Let's say that the bottom of the bathtub is a geopotential surface. The pressure is higher on the deep side because there is more water overhead. The ***pressure gradient*** in the bathtub could be calculated as

$$\text{Pressure gradient} = \frac{\Delta P}{\Delta x} = \frac{P_{\text{shallow}} - P_{\text{deep}}}{1.5\,\text{m}}$$

The pressure gradient is high in the bathtub. Our intuition is that the fluid should flatten out. When that happens, the pressure gradient is zero on the geopotential surface.

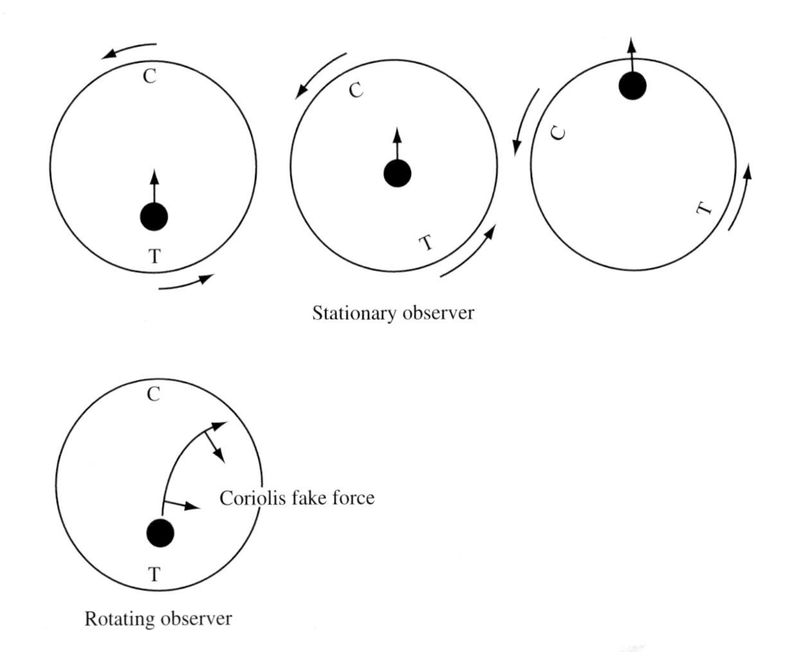

Fig. 6.7 **An illustration of the Coriolis acceleration on a playground merry-go-round. The thrower (T) tosses the ball toward the catcher (C). Before the ball gets to the catcher, the merry-go-round rotates. If we observe this from the point of view of the children on the merry-go-round, it appears as if the ball veers to the right as it moves. The Coriolis acceleration is the kids' attempt to explain the motion of the ball, as if the merry-go-round were not spinning.**

The bathtub is behaving the way Newton's laws say it should; a pressure force pushes fluid, and the fluid begins to flow in the direction the force is pushing.

Flows in the atmosphere or ocean differ from those in the bathtub in that the flow in the former persists for long enough to be steered by the rotation of the Earth. Let's ponder this topic sitting on a merry-go-round at a playground (Fig. 6.7). Two kids sit opposite each other on a merry-go-round tossing a ball back and forth while spinning. The thrower (T) tosses the ball, and it goes straight, as Newton would like. Before the ball reaches the catcher (C) on the other side, the merry-go-round has rotated. The catcher misses the ball. What the kids see is that the ball curved, relative to their rotating world on the merry-go-round.

The same situation applies to a weatherman attempting to describe the motion of the winds on the rotating Earth. The way to interpret a weather map is to think of the Earth as if it were fixed relative to the stars. Then add one other magic force, a fudge force called the ***Coriolis acceleration***, which gives you all the same effects as the rotation of the Earth that we are ignoring. The Coriolis acceleration is a fake force that we imagine is applied to any object in motion. Let's return to the ball on the merry-go-round. Our fake force should change the direction of the ball without changing its speed. To do this, the fake force must be at an angle exactly 90° perpendicular to the direction in which the object is moving. In Fig. 6.7, the Coriolis fake force would have to be directed to the right of the direction of motion, so that it would turn the trajectory of the ball to the right. On Earth, the Coriolis force goes to the right in the northern hemisphere, and to the left in the southern hemisphere.

Now we have to jump from the flat merry-go-round to the gravitating sphere. When we rotate the sphere, what does that do to the people trying to read their weather maps? How much rotation do they feel? The best way to envision this is to imagine a Foucault's pendulum. Foucault pendulums are usually set up in an atrium or stairway so that the weight can hang on a wire 10–20 m long. The wire is so long that the pendulum swings with an enchantingly long, slow period of maybe 10 or 20 s. The weight is rather heavy at 100 kg or so. The wire is mounted on a swivel to allow the pendulum to rotate freely if it chooses to. Once the pendulum has started swinging it goes all day. At the base of the pendulum, some museum employee sets up a circle of little dominoes which the pendulum knocks over as it swings. Over the course of the day, the swing direction of the pendulum changes, knocking over a new little block every hour, on the hour, or some other crowd-pleasing rhythm. Leon Foucault installed the first of these into the Pantheon in Paris for the 1855 Paris Exposition.

Now we can use the pendulum to think about the question of rotation on the surface of the Earth. Let's set up a Foucault pendulum on the North Pole (Fig. 6.8). The Earth spins relative to the fixed stars. The pendulum maintains its swing orientation to be stationary relative to the fixed stars. The pendulum knocks over dominoes on both sides of its swing, so they will all be knocked over after the planet has rotated 180°, which will take 12 h.

Next let's move the pendulum to the equator. The trajectories of the pendulums cannot remain completely motionless with respect to the fixed stars because

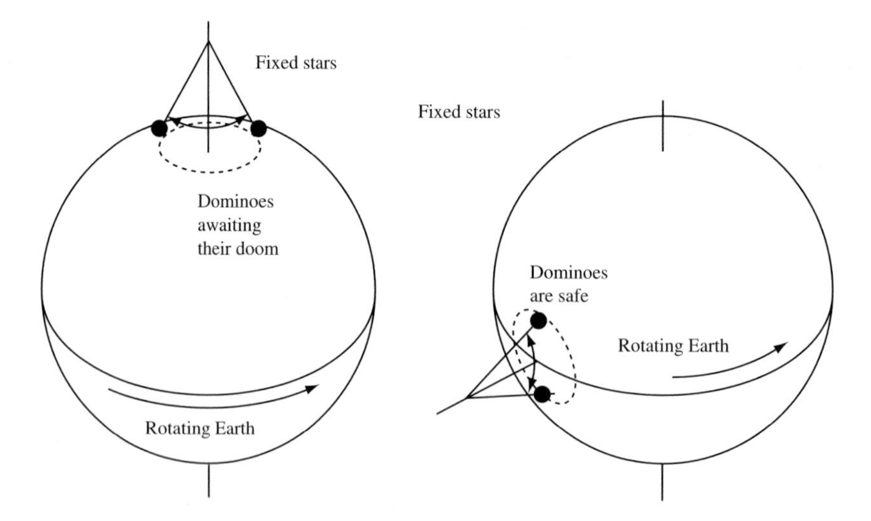

Fig. 6.8 **The Coriolis acceleration that we feel on Earth depends on latitude. A Foucault's pendulum is a rotation detector. A pendulum set in motion at the poles would swing through 180° in 12 h, knocking down all of a circle of dominoes placed around the pendulum. On the equator, the pendulum is started in a north–south direction, and maintains this direction of swing as the Earth rotates. The rate of rotation, and therefore the Coriolis acceleration, is strongest at the poles. At the equator there is no apparent rotation.**

the direction "down" keeps changing as the planet rotates. If we start the pendulum swinging in a north–south direction, the pendulum will keeps its alignment with the stars by continuing to swing north–south. The swing of the pendulum does not appear to rotate to an observer on Earth. The dominoes are safe.

In the middle latitudes, the rate of rotation is in between these extremes of the poles and the equator. The dominoes are completely knocked over in 12 h at the poles, they are never knocked down at the equator. In middle latitudes the knock-down time is longer than 12 h and shorter than forever.

If we release the sloping surface of the water in the bathtub in Fig. 6.6, the water will tend to flow in the direction that the pressure is pushing it, to flow downhill. The driving force for an ocean flow could also be that the wind is blowing (Fig. 6.9). As the fluid begins to flow, the Coriolis acceleration begins to try to deflect the flow to the right (in the northern hemisphere). After a while the fluid is flowing somewhat to the right of the direction that we're pushing it. Eventually, if we wait long enough, the flow will reach a condition called a ***steady state***, a condition in which the flow can persist indefinitely. In the steady state, the Coriolis force balances the driving force, and so there is no net force acting on the fluid. The astonishing implication is that in a rotating world the fluid will eventually end up flowing completely cross-ways to the direction of the forcing! This condition is called ***geostrophic flow***. It is as if the sloping surface in the bathtub would drive a flow across the bathtub. Which direction? Let's stop and figure it out. The pressure from the sloping surface would drive the flow from left to right. An angle of 90° to the right of that would be flowing straight out at the reader, from the page. The water in the bathtub does no such thing, of course, because (i) it doesn't have time to adjust to the steady-state condition in Fig. 6.9, and (ii) the bathtub is not infinitely wide, and water cannot flow through the walls. Great ocean currents that persist for longer than a few days, though, do flow sideways to their driving forces. Sea level on the east side of the Gulf Stream in the North Atlantic is about 1 m higher than it is on the west side. Did I get that right? Figure out the direction that the pressure would be pushing the water, and verify that the flow is 90° to the right.

You can see geostrophic flow on a weather map as cells of high and low pressure with flow going around them (Fig. 6.10). A low-pressure cell in the northern hemisphere has a pressure force pointing inward all around its circumference. At an angle of 90° to the right of that, the winds flow counterclockwise (in the northern hemisphere). Meteorologists call this direction of flow ***cyclonic***. Flow around a high pressure cell is ***anticyclonic***. The fluid flows around the cells rather than into or out of the cells, preserving rather than smoothing out the pressure field that drives it. The ball rolls sideways around the hill rather than directly down the hill as we'd expect. Pressure holes like hurricanes tend to persist in the atmosphere like balls that never run downhill.

Some sets of mathematical equations governing some systems, like the layer model, can be solved directly using algebra to get the exact right answer. The equations of motion for a turbulent fluid are not that easy to solve. We are unable to calculate an exact solution to the equations, but must rather approximate the solution using a computer. A solution we are after consists of temperatures, pressures, flow velocities,

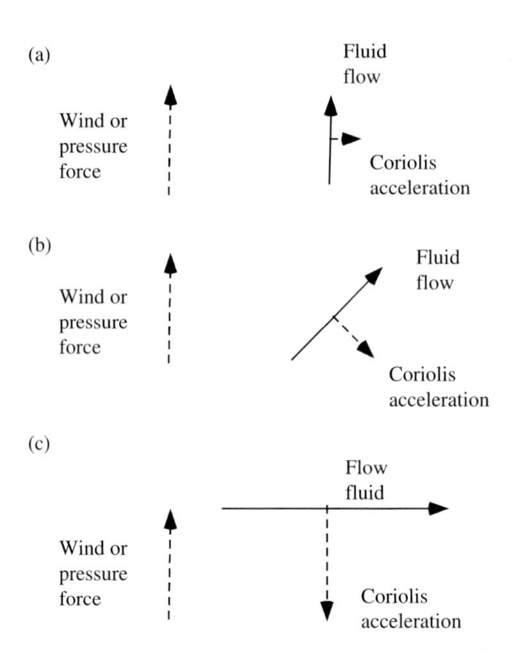

Fig. 6.9 **The Coriolis acceleration affects the way that winds and currents respond to pressure forcing on a rotating planet. In (a, when you first turn on the wind) the fluid initially flows in the direction that the wind or pressure force is pushing it. As it starts flowing, it generates a Coriolis force directed 90° to the direction of its motion. In (b, after a few hours) after a while, the Coriolis force swings the fluid flow toward the right. Eventually, the fluid itself flows 90° to the wind or pressure force, and the Coriolis force just balances the wind or pressure force, in (c, the eventual steady state). This is the steady state, where the flow stops changing and remains steady. In the southern hemisphere, the direction of the Coriolis acceleration and the steady-state flow would be the reverse of that shown here.**

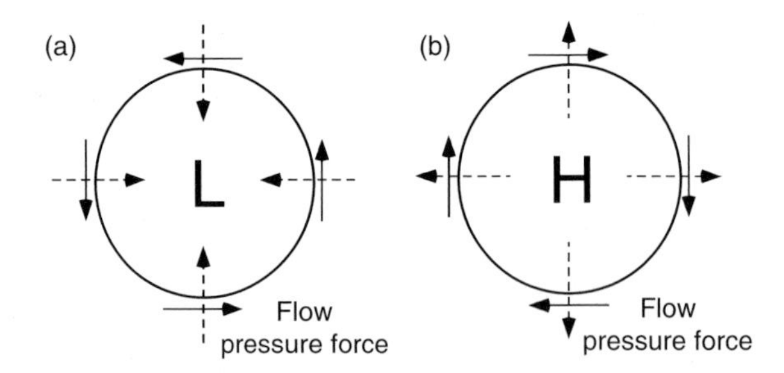

Fig. 6.10 **How pressure variations and the Coriolis acceleration affect weather in the atmosphere. A region of (a) low atmospheric pressure is surrounded by pressure gradients pointing inward, as air tries to flow from high to low pressure (see Fig. 10.6). The steady-state response to that pressure forcing is flow in a direction at 90° to the right of the pressure (see Fig. 6.9), resulting in counterclockwise flow around the pressure hole. The direction of flow would be the opposite around (b) a high-pressure region, and both would be reversed in the southern hemisphere.**

and other quantities related to the flow or the drivers of the flow. These quantities vary from place to place, and as a function of time. To represent them in a ***computer model***, the atmosphere and the ocean are diced up into a three-dimensional ***grid*** of numbers (Fig. 6.11). There is a temperature, a pressure, velocities, and other variables associated with each grid point. The computer model calculates the pressure force at each grid point by comparing the pressures at adjacent grid cells. The model steps forward in time to predict the response of the fluid velocities to the pressure forces. A typical time step for a climate model may be a few minutes. The flows carry around fluid, altering the pressure field, and the simulation snakes it way forward through time. If you would like to play with a climate computer model, you can download one called EdGCM (see Further reading). The model runs on desktop computers and can simulate a few years of simulated time in 24 h of computer time. It has the facility built in to plot maps and time series, such as temperature as a function of time.

Weather simulation has always been a problem ideally suited for computers. Even before computers, researchers like Lewis Richardson dreamed of using human computers to perform the immense numbers of calculations that would be required. Doing the math by hand is so slow, however, that it would never be possible to even keep up with the real weather, let alone make a forecast of the future. ENIAC, one of the first electronic computers constructed in 1945, was used for a range of scientific computation projects including weather forecasting. The fastest computer in the world as of 2005 is called the Earth Simulator, which is in Japan.

Simulation of weather and climate remains one of the grand challenges in the computational sciences. This is because the mechanisms that govern fluid flow often operate at a fairly small spatial scale. If we wanted a model to include everything that governs how cloud drops form, we would have to have grid points every few meters in clouds for example. Fronts in the ocean can be only a few meters wide. To get the answer right, we'd like to have lots of grid points. The trouble is that if we increase the number of grid points by, say, a factor of 10, in each dimension, the total number of grid points goes up by a factor of $10 \times 10 \times 10$ or 1000. That's 1000 times more mathematical operations the computer has to perform in order to do a time step. To make matters worse, as the grid gets more closely spaced, the time step has to get shorter. A grid 10 times finer would require a time step about 10 times shorter. So it would take 10,000 times longer to do a model year of time. State-of-the-art computer models of climate are run at higher resolution all the time, and in general they look more realistic as resolution increases. But they are still far from the resolution they would like, and will be so for the conceivable future.

Some of the physics of the real atmosphere cannot be explicitly resolved, so they must be accounted for by some clever shortcut. The real formation of cloud droplets depends on small-scale turbulent velocities, particle size spectra, and other information which the model is unable to simulate. So the model is programmed to use whatever information it does have, ideally in some intelligent way that might be able to capture some of the behavior of the real world. Cloud formation may be assumed to be some simple function of the humidity of the air, for example, even though we know reality is not so simple. The code word for this approach is ***parameterization***. The humidity of the air is treated as a parameter that controls cloudiness. Other important and

Fig. 6.11 **Surface wind field from a state-of-the-art climate model (a model called FOAM, courtesy of Rob Jacob). This is a snapshot of the winds averaged over one day in the middle of January. The figure is lightly shaded to show sea level pressure, with darker shading indicating lower pressure. One can see the wind blowing counterclockwise around low pressure areas in the northern hemisphere, for example, just to the south of Greenland. In the southern hemisphere, winds blow clockwise around low-pressure areas, for example, in the Southern Ocean south of Australia.**

potentially weak-link parameterizations include the effects of turbulent mixing, air/sea processes such as heat transfer; the list goes on. We will discuss these in more detail in Chapter 11.

Take-home points

1. The energy budget of the Earth fluctuates with time on daily and seasonal timescales (in contrast to the layer model).
2. The annual average energy budget may not balance locally, either, because excess heat from the tropics is carried to high latitudes by winds and ocean currents.
3. The global warming forecast requires simulating the weather, which is a real computational challenge.

Projects

1. *The orbit and seasons.* Answer this question using an online model of the intensity of sunlight as a function of latitude and season at http://understandingtheforecast.org/Projects/orbit.html. The model calculates the distribution of solar heating with latitude and season depending on the orbital parameters of the Earth. Enter a year AD and push calculate. The eccentricity is the extent to which the orbit is out of round; an eccentricity of zero would be a fully circular orbit. Obliquity is the tilt of the Earth's axis of rotation relative to the plane of the Earth's orbit. The third number, the longitude of the vernal equinox, determines the location on the orbit (the date of the year) where northern hemisphere is tilted closest to the Sun. Using the present-day orbital configuration, reproduce Fig. 6.4. Now straighten the tilt of the Earth's axis of rotation by setting obliquity to 0°. What happens to the seasons?

2. *Heat transport.* Answer these questions using an online, full-spectrum radiation model at http://understandingtheforecast.org/Projects/full_spectrum.html.

a. The incoming solar radiation at the equator, averaged over the daily cycle, is about 420 W/m^2. What would the temperature be at the equator if there were no heat transport on Earth? The default albedo in the web interface is 30%, but the real albedo in the tropics may be closer to 10%. What happens then? How much heat transport is required to get a reasonable temperature for the equator? What fraction of the solar influx is this?
b. Repeat the same calculation for high latitudes. Estimate the annual average heat influx at 60° latitude by running the orbital model from the first problem. Just eyeball the fluxes through the year to guess at what the average would be. Now plug this into the full-spectrum light model to see how cold it would be up there if there were no heat transport. If there were no transport and also no storage, how cold would it be in winter?

Further reading

EdGCM, http://edgcm.columbia.edu/ NASA scientists have created a climate GCM model that you can run easily on Windows or on the Macintosh. There are graphical menus for setting up the model run: greenhouse gas concentrations, solar intensity, and so on. There are also graphical tools to make maps and time series of the model output. I can run a few decades of model time overnight on my relatively old Macintosh laptop.

Glieck, James, *Chaos: The Making of a New Science*, Penguin, 1988.

7
Feedbacks

One tricky part about modeling climate is the way that different parts of the climate system interact with each other. Positive feedbacks tend to amplify the variability of climate whereas negative feedbacks provide stability. Water evaporates into warmer air, acting as a greenhouse gas to amplify the initial warming. Melting ice changes the albedo of the Earth, acting as a positive feedback to temperature variations. Clouds have a huge influence on climate by interacting with both visible and IR light. We can point to present-day human influence on cloudiness, but it is difficult to predict how clouds will feed back to long-term climate change. Ocean circulation has feedback interactions with climate, in the equatorial Pacific and in the North Atlantic. In general, real climate variations appear to be larger than our models predict because it is difficult to capture all of the feedbacks in models.

Positive and negative feedbacks

The challenging thing about modeling climate is that there are ***feedbacks***. A feedback is a loop of cause and effect (Fig. 7.1). It will be easier to wade through this particular thicket if we make a few abstruse-sounding definitions here. The temperature is the quantity that gets fed-back in many examples, although feedbacks also control populations, and chemistry and water vapor lots of other things. So we can be general: let's call the quantity that is being fed-back the ***state variable***. The loop of causality is called the ***feedback loop***. The arrow coming in from stage left is an ***input perturbation***.

One example of a feedback loop is the ***ice albedo feedback***, operating on the state variable of temperature. An input perturbation, such as a rise in greenhouse gases, drives temperature up a bit. Ice melts, reducing the albedo (the reflectivity of the Earth: Chapter 3), which drives temperature up a bit more. This is called a ***positive feedback*** because the directions of the input perturbation and the feedback loop agree with each other. A positive feedback can work in the opposite direction as well, during glacial time for example: colder, more ice, colder still. The point is that the feedback loop, in this case, reinforces the direction of change in the input. A positive feedback loop is an amplifier.

Negative feedbacks tend to stabilize things. One example we have already discussed is the way that the Earth regulates its heat flux to space, according to the Stefan–Boltzmann rule σT^4. If the Earth is too warm, the outgoing energy flux will be greater than the incoming solar energy flux, and the Earth will cool. The water vapor concentration in the atmosphere at a particular temperature, which we will run into in the next section, is another example.

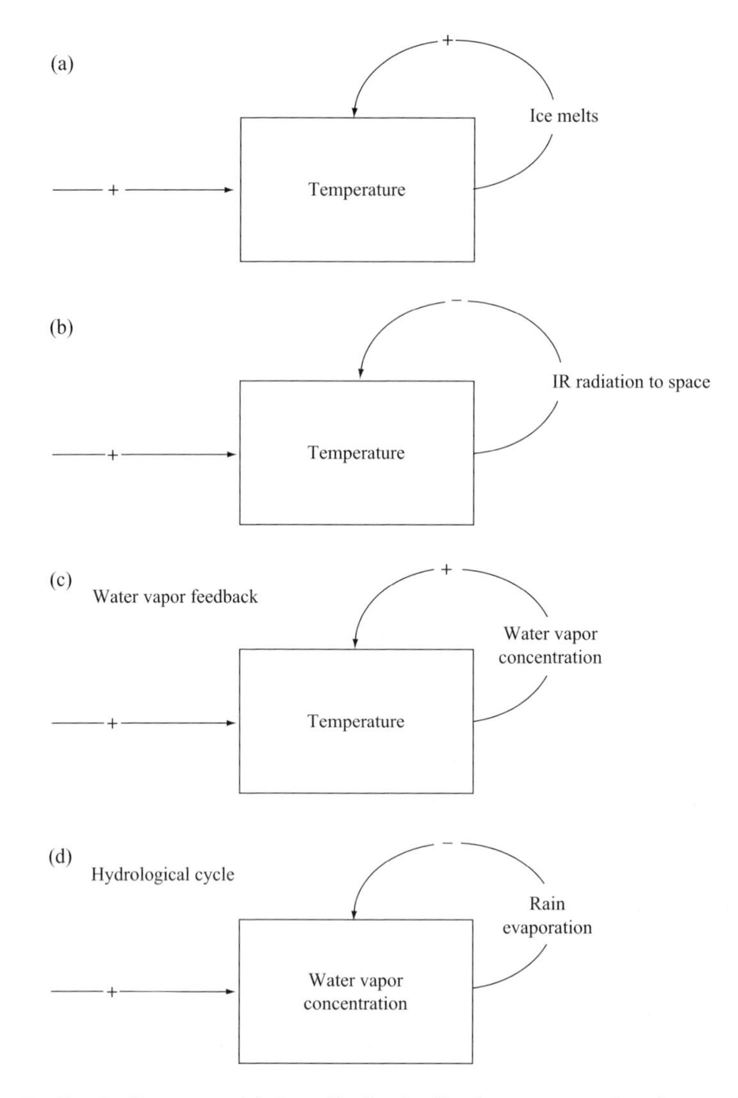

Fig. 7.1 **Feedback diagrams. (a) Ice albedo feedback: an example of a positive feedback. Some external perturbation increases the temperature. The increase in temperature causes ice to melt, allowing the land to absorb more of the incoming solar radiation (by decreasing the albedo). The melting ice drives temperature up further. (b) Stefan–Boltzmann feedback: an example of a negative feedback, resulting from the Stefan–Boltzmann IR energy flux σT^4. (c) is the water vapor feedback, which amplifies the temperature effect of rising CO_2. (d) Hydrological cycle: the negative feedback that controls atmospheric water vapor, given an atmospheric temperature.**

Water vapor feedback

Water vapor is responsible for more greenhouse heat trapping on Earth than CO_2 is, and yet your impression is that global warming is primarily about CO_2, not water vapor.

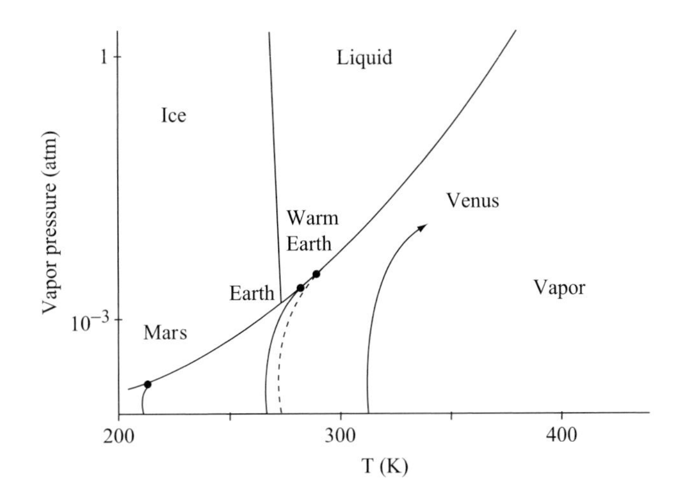

Fig. 7.2 **A phase diagram for water demonstrating that the water vapor feedback on Earth and Mars is limited, while Venus is free to have a runaway greenhouse effect.**

No one screams about global warming when you run your lawn sprinkler. Why not? The answer is that, at some given temperature, the amount of water vapor is controlled by negative feedbacks with the rate of evaporation and with rain. Our lawn sprinkler may be depleting our fresh water supply, but it is not going to lead to global warming because any water we drive into the atmosphere will just rain out next week. A negative feedback stabilizes the water vapor concentration at any given temperature (Fig. 7.1d).

The positive feedback arises when we start to think about water vapor affecting a state variable of temperature (Fig. 7.1c). An initial increase in temperature will allow more water to evaporate (Fig. 5.6). Water vapor is a greenhouse gas, so its increase tends to warm the planet still further.

Where does this end, you may wonder to yourself. Does a positive feedback ever stop, or does it explode? The water vapor feedback can in principle; this scenario is called a ***runaway greenhouse effect.*** It happened on Venus, but we are in no immediate danger of it happening on Earth. The easiest way to visualize the water vapor feedback stopping or not stopping is to draw water vapor feedback trajectories on a phase diagram for water (Fig. 7.2). The ***phases*** of water are solid, liquid, and vapor. The phase diagram for water shows us which phase or phases of water you will find, if you subject water to some pressure and temperature. At high temperature and low pressure, you expect to find vapor. Vapor likes it hot. As you decrease the temperature and increase the pressure, you get liquid, and when it gets cold enough you get ice. On the boundaries between the regions you get two phases, like vapor and water, or ice and water. At the triple point, a certain pressure and temperature combination, you get all three phases coexisting together, a glass of boiling ice water. Of course, you can throw ice cubes in a pot of boiling water on your stove any time you like, but the ice will quickly melt because it is not at equilibrium. The phase plot in Fig. 7.2 represents equilibrium conditions.

To understand the water vapor feedback, we are going to imagine what happens if we suddenly introduce water to a planet that initially had no water. The starting position on

Fig. 7.2 is all the way at the bottom of the figure, where the pressure of water vapor in the air would be low. From this starting point, what will happen is that water will evaporate, and the vapor pressure in the atmosphere will rise. Water vapor is a greenhouse gas, so as the water vapor content of the atmosphere increases, the temperature warms. Therefore, as the condition of the planet moves upward on the figure (higher water vapor pressure), it also moves somewhat to the right (higher temperature).

The middle curve on Fig. 7.2, labeled Earth, moves up and to the right until it intersects the stability field of water. At this point, the atmosphere is holding as much water vapor as it can carry. Any further evaporation just makes it rain. The water vapor feedback has increased the temperature of the planet above what it would have been if it were dry, but the positive feedback did not run away as it was limited by a tendency to rain. The curve on the left on Fig. 7.2 represents something like the situation on Mars. Here the water vapor feedback path intersects the stability field of ice. The water vapor feedback does not lead to a runaway greenhouse effect on Mars or Earth because the feedback is limited by the stability field of water or ice.

We only see the runaway greenhouse effect in the path on the right, labeled Venus. If water were introduced on Venus it would evaporate, increasing the temperature, as shown. The difference here is that this path never intersects the stability fields of either liquid or solid water. Planetary scientists presume that Venus originally had about as much water as Earth, but the high solar heat flux associated with orbiting so close to the Sun forced that water to evaporate into the atmosphere, rather than condense into oceans as it has on Earth.

Evaporation of a planet's water is a one-way street because if water vapor reaches the upper atmosphere, its chemical bonds will get blown apart by the intense ultraviolet (UV) light in the upper atmosphere. The hydrogen atoms, once they are disconnected from oxygen, are small enough and light enough that they are able to escape into space. The water is lost for good. This is the presumed fate of Venus' water, but Earth has retained its water because the air is so cold in the tropopause, the layer of atmosphere that separates the troposphere from the stratosphere (Fig. 5.1). The troposphere contains lots of water, but the stratosphere is dry because of the cold trap of the tropopause. The oceans, viewed from space, look heartbreakingly vulnerable, but they are protected, and apparently have been for billions of years, by a thin layer of air. Marvelous!

Theoretically, if the Sun were to get hotter, or if CO_2 concentrations were high enough, the Earth could move to the right on Fig. 7.2, sufficiently far that it could escape the liquid water stability field and hence run away. But don't worry, there is not enough fossil fuel carbon on Earth to do this. The Sun is heating up over geologic time, but only very gradually, and the runaway greenhouse effect from this is nothing to worry about for billions of years. Interestingly, if the equator were isolated from the poles, blocked from any heat transport and forced to balance its energy fluxes using outgoing IR only, the tropics would be a runaway greenhouse. It is a lucky thing that we have the high latitudes to act as cooling fins!

The role that the water vapor feedback plays in the global warming forecast can be drawn on Fig. 7.2, although global warming climate change is tiny compared with Venus' climate, so I've drawn it horribly exaggerated so we can see it. An initial

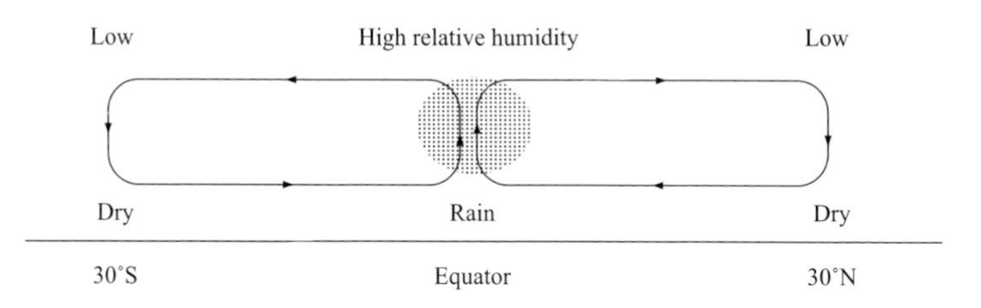

Fig. 7.3 **The Hadley circulation and its effect on the humidity of the air in the troposphere.**

temperature perturbation, say from fossil fuel CO_2 release, moves the water vapor feedback path a bit toward the right, toward Venus. The path ends as it hits the liquid saturation line, as for the unperturbed Earth but at a higher temperature. The temperature difference between the two raining, moist Earths (where they intersect liquid water) is greater than the temperature difference between the two dry Earths (at the bottom of the plot). The water vapor feedback amplifies the temperature change that you would get from increasing CO_2 on a dry planet.

On the real Earth, mercifully, the relative humidity is not everywhere 100%. One mechanism which controls a lot of the atmospheric water vapor franchise is called ***Hadley circulation*** (Fig. 7.3). Warm air at the equator rises convectively. Water condenses as the air rises and cools. This column of air has a lot of water vapor in it, in general. The air spreads out at high altitude, then begins to ***subside*** in the subtropics, about 30° latitude north and south. That air has been through the wringer, the cold tropopause, and there is not much water vapor left in it. The great deserts of the world are located under these dry air blowers. The air flows back equatorward along the surface, picking up water vapor as it goes. Globally, the average humidity of surface air is about 80%.

One could imagine changing the winds, the Hadley circulation for example, and changing the humidity of the surface air. Maybe if the Hadley circulation were stronger, the surface air would dry out a bit. Could a mechanism like this counteract the warming effect of rising CO_2? This is a difficult question to answer for sure because it hinges on issues of turbulence and fluid flow, which are impossible to model perfectly (Chapter 6). Models predictions tend to agree, however, that the amount of water vapor in the atmosphere tends to increase with warming from rising CO_2 because warm air holds more water vapor than cold (Fig. 5.6). The water vapor feedback is predicted to be extremely important for future climate change, doubling or tripling the warming that we would expect if Earth were a dry planet. If it weren't for the water vapor feedback, maybe there would be no need to worry about rising CO_2 concentrations!

Ice albedo feedback

The ice albedo feedback is a positive one that operates mostly in the high latitudes because that is where the ice is. According to our current climate models, global temperature changes will be amplified in the high latitudes, by a factor of about 2 or 4 over global average warming. Climate change today is much more obvious in high latitudes, as in melting of permafrost in Alaska and Siberia and melting of sea ice in the Arctic, than it is in other places like the tropics, for this reason.

In addition to changing the albedo, the presence or absence of ice on the ocean has a huge effect on the local air temperature. Air over water doesn't get much colder than freezing, 0°C, but air over ice or land can be much colder. As we learned in Chapter 6, water absorbs and emits heat, buffering the extremes of temperature, whereas land and ice do not so much. Sea ice changes may have a lot to do with fast, extreme changes in climate reflected in Greenland ice cores during glacial climate.

Clouds

Clouds are the sleeping giant of the climate system. Most of the variation in sensitivities of different climate models comes from clouds. Clouds interact and interfere with both visible and IR light. It turns out that according to visible light, clouds should cool the planet while with IR light clouds should warm the planet. It gets even more complicated; which of these two effects wins depends on the altitude of the clouds. High-altitude clouds don't reflect much visible light, but they are pure hell in the IR. Lower clouds are the reverse; less of an IR effect because they are warm, but a large change in albedo because they are thick and opaque.

You have already encountered the IR side of the story, in Chapter 3. Clouds are tolerably good blackbodies, emitting a nice clean blackbody spectrum of light at whatever temperature it finds itself. Light escapes to space from the tops of the clouds, so the crucial thing to know is how high the tops of the clouds are, and therefore what the ***cloud top temperature*** is. High clouds emit IR light at cold temperatures blocking the bright IR light from the warm ground. Clouds with low tops have a smaller effect on the outgoing IR light.

Clouds also reflect incoming solar radiation, increasing the albedo of the Earth. The electric field of incoming visible light stimulates the electric field in the water droplet to oscillate, and this oscillation emits light of its own, at the same frequency as the incoming light. The name for this process is ***scattering***. This interaction of light with matter is not the same as light absorption because the energy never gets converted to thermal energy of the water. Absorbed light would be reradiated in the IR, but scattered light in effect bounces off the water droplets as photons at the same frequency they rode in on. Some of the light energy is absorbed also, and the fraction of scattered versus absorbed light depends on many things. As a result, the albedo of clouds varies widely between different types of clouds. Scientists are struggling to figure out how real clouds work and how they might be changing already. Modeling them perfectly is currently out of our reach. The best we can do is a very crude approximation.

One important factor in determining the albedo of a cloud is the amount of water it contains. Low clouds contain more water than high clouds, in general. Clouds that are about to rain have more water in them than nonraining clouds. Another factor is the size of the cloud drops. Scattering is more efficient in smaller drops if the drop size is smaller. A third factor is the directionality of the scattered light. Most of the light scattered by spherical liquid drops continue in a more or less forward direction, which, if the light is coming down vertically, would mean to continue downward toward the Earth. Ice crystals are better at actually reversing the direction of light, sending it back up into space, increasing the albedo.

As air rises it cools and the ***saturation vapor pressure*** decreases (Fig. 5.6). We heard about this process in Chapter 5, but let's look at the cloud formation process in a little more detail. As soon as the relative humidity reaches 100%, there would be an energy payback if the water were to condense. But it might not happen right away. The water can get "stuck" supersaturated, that is, humidity can exceed 100%, by quite a bit, if the air is very clean with no particles in it for the droplets to form around. Liquid water has the property that it abhors the gas/liquid surface. Abhors may be a strong word, but the water molecules have a definite preference to be immersed, with other liquid water molecules on all sides, rather than sitting on the surface of the drop exposed to the air. Small particles have a greater proportion of these unhappy, unfulfilled surface molecules. Therefore, the water in small droplets has a slightly higher tendency to evaporate into the gas phase. When we report relative humidity, we are reporting it relative to water that has a normal, flat surface, or maybe large droplets. The saturation vapor pressure in equilibrium with small droplets is higher than it is for lower-energy water molecules with a flat surface at the same temperature. We will require a relative humidity higher than 100% if we wish to form small cloud droplets.

Most air in the troposphere has particles in it before any cloud drops start to form, around which the water droplets may begin to grow. These particles are known as ***cloud condensation nuclei***. Clouds begin to form at lower relative humidity if condensation nuclei are present. The condensation nuclei are of the order of 0.1 μm in size. Once the droplets form, the large droplets in a cloud tend to grow at the expense of the small ones because the latter have a greater tendency to evaporate. Urbanization of cloud droplets. The droplets in clouds tend to grow fairly quickly to sizes of around 5 μm, larger in clouds that are about to rain. At high altitude, ice has a tendency to form, either by freezing liquid or by forming directly from the vapor. Vapor deposition releases latent heat on a growing ice crystal as it grows, resulting in the amazing symmetric patterns that snowflakes are formed in. No two alike, they say. That would be because no two snowflakes have exactly the same temperature and water vapor supersaturation histories.

There are three main types of clouds (Fig. 7.4). Cirrus clouds are located up at 8–12 km altitude. These are the only types of clouds up that high. Between cirrus clouds and low-altitude clouds there are several kilometers of air space where you typically do not find clouds. Cirrus clouds contain 10 or 100 times less water per volume than lower altitude clouds typically hold. Because they are so thin, cirrus clouds do not block incoming solar radiation as effectively as lower clouds do. You can often see blue sky right through cirrus clouds. The albedo impact of cirrus clouds is therefore weaker than it is for lower, thicker clouds. The IR effect of cirrus clouds is strong because the cloud is effective enough as a blackbody to block intense IR from the warm ground, substituting for it weaker IR from the cold upper atmosphere.

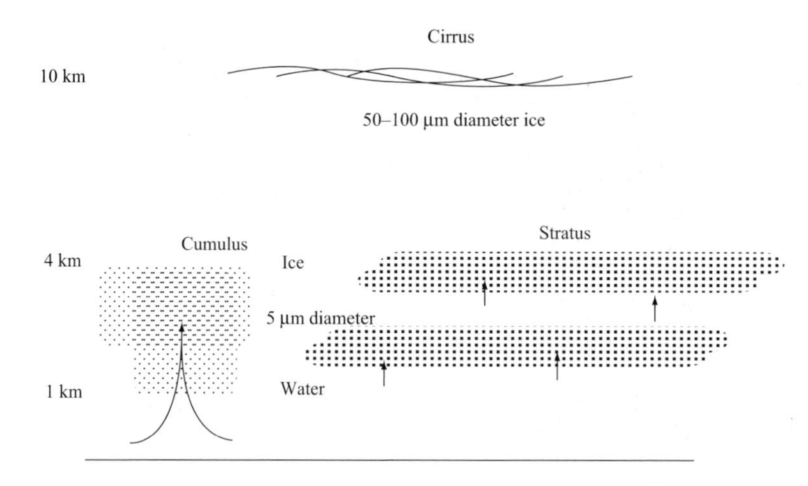

Fig. 7.4 **The three main types of clouds.**

The two main types of clouds at low altitude are called ***cumulus*** and ***stratus***. Cumulus clouds are towers, the result of a focused blast of convection. Thunderstorms come from cumulus clouds. Stratus clouds are layered, formed by broad diffuse upward motion spread out over a large geographical area. Both types of fluid motion are difficult to simulate in climate models; the focused motions because they occur on spatial scales that are smaller than the grid sizes of the models, and the broad diffusive upward velocities because they are so slow and are difficult to predict accurately. Stratus clouds are particularly important to get right for the climate forecast because they affect the albedo of such a large area of the Earth's surface. Two-thirds of the albedo of the Earth derives from clouds, stratus clouds in large measure.

There is plenty of scope for human activity to impact the climate by altering clouds. Anthropogenic condensation nuclei are mostly combustion products. The abundance of cirrus clouds on Earth may be augmented by the ***contrails*** left by jet airplanes (Fig 7.5). In clean air, in the absence of cloud condensation nuclei, the air can be supersaturated indefinitely without forming clouds. An airplane passes through and creates a trail of air with higher humidity, from combusting the hydrogen in the jet fuel and with exhaust particles that can serve as condensation nuclei. Droplets form very quickly behind the airplane and then continue to grow and to spread out. Most of the moisture in the contrail comes from the ambient air, not from the airplane. The cloud particles spread out and eventually become indistinguishable from natural cirrus particles. For this reason it is difficult to know how much impact aircraft have on the cloudiness of the upper atmosphere. Contrails tend to warm the planet, by perhaps a few percentage of the total human-induced warming. A difference between contrail warming and CO_2 warming is of course that CO_2 accumulates in the atmosphere whereas contrails would dissipate in a few days, if we stopped creating them.

Low-altitude clouds can be affected by smoke particles from coal power plants, in particular, sulfur in coal. Sulfur is emitted in flue gas as SO_2 and it oxidizes in a week or so in the atmosphere to sulfuric acid, H_2SO_4. The sulfuric acid condenses into small droplets less than 1 μm in size called ***sulfate aerosols***. Eventually, the sulfuric acid

Fig. 7.5 **Contrails.**

rains out and is called ***acid rain***. Sulfate aerosols are extremely good cloud condensation nuclei because the strong acid tends to pull water out of the air like those silica gel packets that are shipped with electronics equipment. Natural condensation nuclei include sea salt, dust, pollen, smoke, and sulfur compounds emitted by phytoplankton. Most parts of the world have enough natural condensation nuclei so the issue is not whether to form a droplet or not, the way it is for cirrus clouds and contrails. Rather, the issue is the size of the droplets. If there are plenty of condensation nuclei around, more droplets will form, and the average drop will be smaller. This is a big deal because smaller drops scatter more efficiently and therefore have a higher albedo. Get this: a change in droplet size from 10 μm down to 8 μm has the same effect on the radiation balance of the Earth as doubling the CO_2 concentration. That's a big deal from such tiny little droplets! The potential for sulfate aerosols to change the albedo of the Earth by changing the average size of cloud droplets is called the ***sulfate aerosol indirect effect***. This is a potentially huge effect, but the uncertainty is as large as the estimated effect (Fig. 10.2). One indication that this might be a real effect comes from ship tracks in the ocean, which are followed by clouds of lower-atmosphere droplets just like contrails (Fig 7.6).

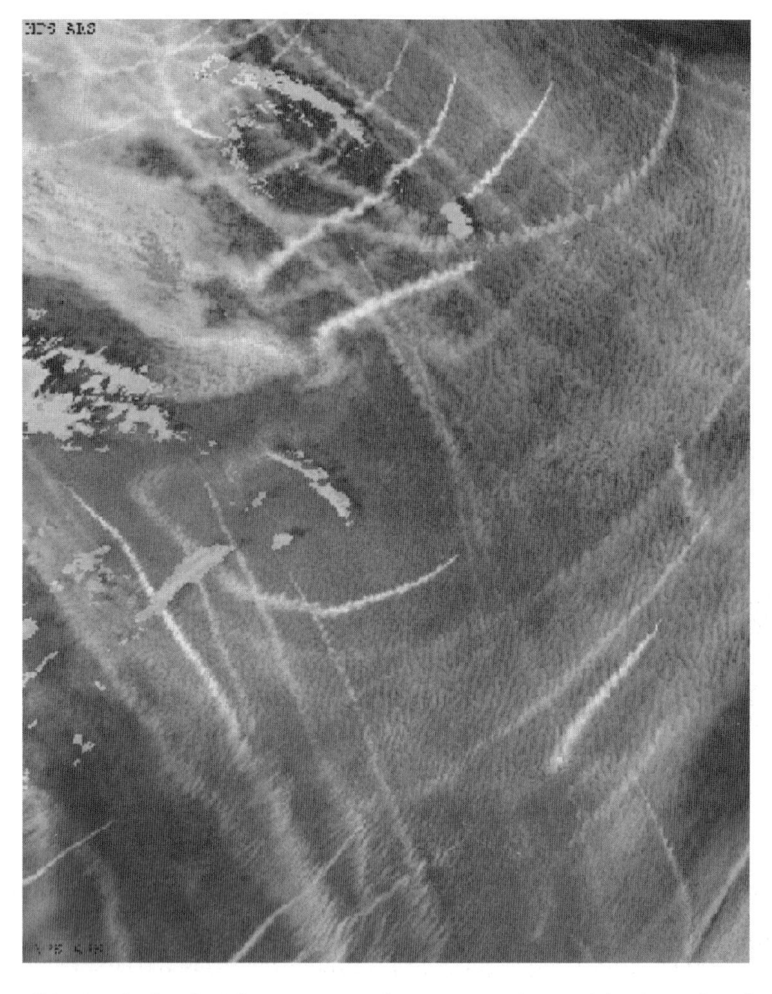

Fig. 7.6 **Ship track clouds. These suggest that man-made particles introduced to clean marine air may nucleate cloud droplets.**

Ocean currents

The oceans interact with climate, in many ways. One example is called the ***el Niño*** oscillation, a flip-flop between two self-reinforcing states, el Niño and la Niña (Fig. 7.7). During la Niña, the thermocline along the equator in the Pacific Ocean slopes upward toward the east, bringing cold waters to the sea surface. The temperature contrast between east and west drives winds which travel from east to west, maintaining the slope of the thermocline. The other state, el Niño, flattens the slope in the thermocline and collapses the wind. The coupled atmosphere/ocean system flips back and forth between these two climate states one cycle every 4–7 years. The state of la Niña affects climate to varying extents all around the world. There is a possibility that with global warming, the Pacific may tend to favor the la Niña state, but the model forecasts for this are not very reliable.

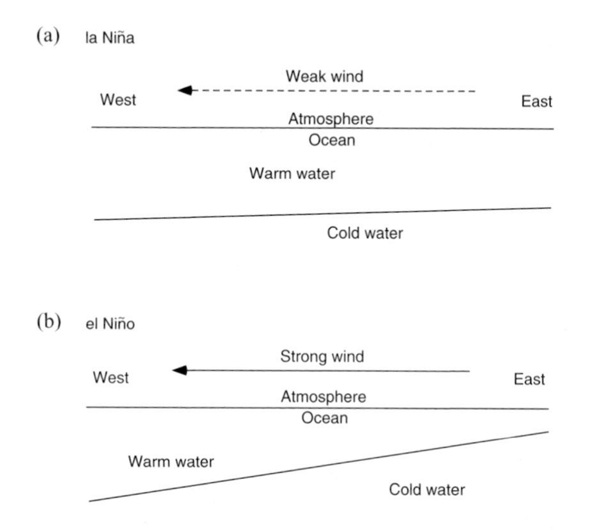

Fig. 7.7 **The configuration of the atmosphere and ocean along the equator in the Pacific Ocean during el Niño and la Niña phases of the el Niño climate oscillation. (a) la Niña: In the la Niña phase of the oscillation, the slope in the ocean temperature boundary, and the winds in the atmosphere, collapse. The climate oscillates back and forth between these two states every 4–7 years or so, affecting climate conditions around the rest of the world. (b) el Niño: During el Niño, the boundary between warm water and cold water in the ocean slopes up from west to east, exposing cold water to the atmosphere in the east near Peru. In addition to stimulating the Peruvian fisheries, the cold water drives a strong wind from east to west which tends to sustain the slope in the ocean temperature boundary.**

Another potential feedback from the ocean to climate is in the North Atlantic Ocean. Warm water is carried to the North Atlantic in the Gulf Stream discovered by Benjamin Franklin. As the water cools, its density increases and it sinks to the deep Atlantic, making room for more warm water to carry more heat to the high northern latitudes. Our climatic and oceanographic reconstructions of the North Atlantic region from the last ice age point to an instability in this system. The climate record in the ice in Greenland shows huge temperature swings called Dansgaard-Oeschger events – temperature changes of 10°C at that location within a few years. Oceanographic reconstructions from sediment cores show that the overturning circulation in the North Atlantic starts and stops in synchrony with these temperature swings. One large climate shift, 8200 years ago, has been correlated with the catastrophic release of a Great Lake's worth of fresh water into the North Atlantic, as an ice dam holding the lake in finally burst. Fresh water is not as dense as salty water, so an input of fresh water could have stopped the overturning circulation.

The circulation in the North Atlantic may change in the coming century in response to rising temperature or precipitation. Warming is more intense in high latitudes than the global average because of the ice albedo feedback. Warming tends to increase rain and snowfall because there is more water vapor in warm air than cold air. Another potential source of new fresh water to the North Atlantic is by melting of the Greenland ice sheet if it were to melt quickly enough.

Terrestrial biosphere feedbacks

The terrestrial biosphere has the potential to feed back to climate by altering the albedo of the land, and by altering evaporation. Trees have a much lower albedo than does the bare ground. Cooling might lead to the demise of forests, increasing albedo, and driving temperatures down further. Trees also mine ground water and evaporate it from their leaves. The Amazon rain forest is said to perpetuate its own existence by extracting ground water and evaporating it to the air, rather than allowing it to escape through rivers to the ocean. By recycling the water to the atmosphere, the forest increases the rate of rainfall that keeps the rain forest healthy.

Carbon cycle feedbacks

There are many feedbacks within the carbon cycle, which is the subject of the next three chapters. Warming the atmosphere increases the temperature of the surface ocean, which tends to drive CO_2 to degas from the ocean to the atmosphere. The ice ages are thought to be ultimately triggered by variations in the Earth's orbit, but the climate shifts would have been much smaller if the ocean hadn't contrived to pull CO_2 out of the atmosphere as the ice sheets started to grow. The biosphere on land, trees, soils, and so on may be stimulated to take up or release CO_2 as climate changes. The atmospheric methane concentration may increase as a result of climate change, with melting permafrost and creation of swamps by increased rainfall.

Feedbacks in the paleoclimate record

In general, first-attempt climate models seem to have an inherent tendency to underpredict the extremes of climate variation in the real climate, as inferred from climate records such as ice cores. I believe this is due to the difficulty of capturing feedbacks in climate models. Feedbacks often couple together very different parts of the climate system, requiring a cloud physicist to work together with an oceanographer, for example. It takes creativity to think of some of these relationships! The message is that the future may surprise us and test our creativity in thinking of couplings, interactions, and feedbacks, that may not be obvious at first glance.

Take-home points

1. Positive feedbacks act as amplifiers of variability whereas negative feedbacks act as stabilizers.
2. The water vapor feedback doubles or triples the expected warming due to rising CO_2 concentrations.
3. The ice albedo feedback amplifies the warming in high latitudes by a factor of three or four.

4. Clouds have a huge impact on climate. Human activity is already changing the nature of clouds on Earth. Future changes are difficult to predict.

Projects

Answer these questions using the full-spectrum radiation model at http://understanding theforecast/Projects/full_spectrum.html.

1. *Compare two codes.* You will find that the two radiation codes give very different answers for the temperature sensitivity to CO_2 and water vapor. What is the ΔT_{2x} for each model? Is it the same for doubling from 100 to 200 ppm as it is for 350 to 700 ppm? The model includes the water vapor feedback automatically, but we can turn this off by zeroing the relative humidity. What is the ΔT_{2x} without the water vapor feedback? How comparable are the CCM3 and Chou models?

2. *Clouds and full-spectrum light.* Let's look at the effects of clouds. For each radiation code, document the effect of, say, a 100% cloud cover, for high and low clouds (run each separately). Which type of cloud has the largest effect?

   a. What is the effect of changing the drop size from 10 to 8 μm in the low clouds? How do these radiative effects compare with the effect of doubling CO_2?
   b. Run a set of scenarios with high clouds, using a range of water contents, from 100 down to 2 g/m^2. Make a plot of the equilibrium temperature depending on the water content. You should get a fairly surprising looking plot. Can you explain what you see?

3. *Clouds and upwelling IR light.*
Answer the next two questions using http://understandingtheforecast/Projects/ infrared_spectrum.html. What is the effect of clouds on the outgoing IR energy flux of the atmosphere?

   a. Are higher clouds or lower clouds the most significant to the outgoing IR energy balance?
   b. Can you see the effect of the clouds in the outgoing spectra? How is it that clouds change the outgoing IR flux?

4. *Clouds and downwelling IR.* Set the Sensor Altitude to 0 km, and choose the Looking Up option at the bottom of the model web page. Do this with no clouds, and then again with clouds. Explain what you see. Why, at night, is it warmer when there are clouds?

Further reading

Kump, Lee R., James F. Kasting, and Robert G. Crane, *The Earth System*, Prentice-Hall, 1999.
Walker, James C.G., *Numerical Adventures with Geochemical Models*, Oxford Press, 1991.

Part II

The carbon cycle

8
Carbon on Earth

Carbon exists in a range of oxidation states and chemical forms. Carbon is most stable in the oxidized form, on Earth mostly as CO_2 and $CaCO_3$. Photosynthesis converts carbon to higher-energy reduced form, both to store energy from the Sun and because the versatility of reduced carbon chemistry makes a natural building block for life.

The atmosphere contains only a tiny fraction of the carbon on Earth. The terrestrial biosphere has several times more carbon, if soil carbon is included in the reckoning. The seasonal breathing of the terrestrial biosphere is measurable in a seasonal cycle of atmospheric CO_2. The ocean contains 50 times as much carbon as the atmosphere and is apparently responsible for large changes in atmospheric CO_2 over 100,000-year glacial cycles, although no one is quite sure how. On timescales of millions of years, the weathering of igneous rocks consumes CO_2, stabilizing atmospheric CO_2 and the climate of the Earth in the process.

The chemistry of carbon

Carbon has a rich, spectacular, wondrous chemistry. There are more scientists studying the chemistry of carbon than any other element, I am sure. The nearest relative of carbon is the element silicon which is located directly underneath carbon on the periodic table. Silicon is one of the more abundant elements on Earth, and the chemistry of the rocks in the mantle is controlled to a large extent by the whims of silicon. Silicon tossed out into the environment finds its way into hundreds of different crystal forms, including the minerals in rocks and clays. However, silicon chemistry is regimented, flat, one could almost say lifeless, compared with the chemistry of carbon. Carbon chemistry is kept highly organized within living things, but after life is finished with it, carbon left in soils, sediments, and rocks forms itself into an indescribable goo called ***humic acids*** or ***kerogen***. Silicon forms hundreds of crystals, but the number of possible configurations of carbon is essentially infinite. I scoff at the cliché from *Star Trek*, "carbon-based life forms" spoken as though this were some sort of novel idea. Perhaps I am close-minded, lacking vision, but I have difficulty imagining life based primarily on any element other than carbon.

Planetary scientists presume that the three terrestrial planets Venus, Earth, and Mars all received about the same amounts of carbon when they formed. The carbon on Venus is mostly found in the atmosphere, and there is so much carbon that the pressure on the surface of Venus is 70 times higher than the total atmospheric pressure on Earth. The carbon on Earth is mostly found in limestone sedimentary rocks, with only 0.38% of

1 atm of pressure. The amount of CO_2 in the atmosphere of Venus is therefore 180,000 times higher than the amount in Earth's atmosphere. Clearly a planet has a wide range of options about where to put its carbon.

Carbon on Earth is the backbone for constructing the machinery of life. In addition to serving as the framework and scaffolding for that machinery, carbon chemistry also provides the means of storing energy. Photosynthesis converts the energy of sunlight into carbon biomolecules. This energy might be used by the plant, or perhaps an animal eats the plant and uses its energy for itself. Over geologic timescales, carbon stores energy on a planetary scale, "charging up" the biosphere with a store of biomolecules and atmospheric oxygen like a gigantic battery. When we extract energy from fossil fuels, we harvest some of the ancient energy stores of the biosphere. In the process we rearrange the distribution of carbon among its various reservoirs on Earth.

To understand how all of this works, we need to know about a fundamental property of atoms in molecules, called ***oxidation state***. An example of oxidation state that you are familiar with is that iron oxidizes when it rusts. Three simple carbon molecules demonstrate the spectrum of reduction/oxidation (abbreviated ***redox***) chemistry for carbon.

	Oxidized	Intermediate	Reduced
Simplest example	CO_2	CH_2O	CH_4
Carbon oxidation state	+4	0	−4
General category	Inorganic carbon	Carbohydrates	Hydrocarbons

The first is carbon dioxide, which we learned in Chapter 3 is a gas, a greenhouse gas at that. The last is methane, also a greenhouse gas. CO_2 is the ***oxidized*** form of carbon in this spectrum, and methane is the ***reduced*** form. You might guess that the phrase "oxidation state" implies that we must be talking about oxygen in some way. In more general terms, oxidation state is a measure of the surplus or deficit of ***electrons*** around the carbon. Oxygen is a very greedy element for the two electrons that it requires to find its most stable electronic configuration. Usually oxygen in a molecule is credited with stealing two electrons from its bond partner (an exception is if its bond partner is another oxygen, in which case they must share equally). The carbon in CO_2 has two bonds with oxygen, so it has given up four electrons, two to each oxygen atom. Since each electron has a charge of −1, a deficit of four electrons leaves the carbon in CO_2 with an oxidation state of +4. Hydrogen, in contrast, is relatively generous with its single electron. The four hydrogens in methane each donate one electron to carbon, so carbon ends up with an oxidation state of −4. The oxidation state of CH_2O, the middle compound in the spectrum, is zero, because carbon gains two electrons from the hydrogens, but donates two to the oxygen.

There are many other carbon compounds that have these oxidation states, and it makes sense to group them together into families in this way. Most of the carbon on Earth is oxidized carbon, also called ***inorganic carbon***, in $CaCO_3$ (limestone) rocks or dissolved in the oceans. Inorganic carbon has a fascinating chemistry (spoken as one who has devoted his life to understanding it), but it is nowhere near complex enough to create the machinery of life. At the reduced end of the spectrum, ***hydrocarbons*** include oil as well as natural gas. Some biomolecules are hydrocarbons, such as fats. Again, amazing stuff, but life could not be constructed from fats alone. Life is comprised of carbon largely in the intermediate oxidation state called ***carbohydrates***. Sugars for example have formulas that are multiples of that of humble CH_2O (a chemical called formaldehyde), like glucose for example is $(CH_2O)_6$. Hydrocarbons and carbohydrates are together referred to as ***organic carbon***. Here is a case where scientific and common usages of a word diverge strikingly. To a chemist, compounds made of reduced carbon, such as DDT, dioxin, and PCBs, are considered organic compounds. To the common person, "organic" produce at the grocery store is supposed to be free of toxic manmade compounds such as these.

The stablest form of carbon in the chemical environment of the Earth is the oxidized form. The carbon pools on Venus and Mars are both nearly entirely oxidized. Life on Earth is based on the nifty trick of harvesting energy from sunlight and storing it by creating reduced carbon from oxidized carbon

$$CO_2 + H_2O + \text{Energy} \Leftrightarrow CH_2O + O_2 \qquad (8.1)$$

The forward direction of reaction (8.1) is called ***photosynthesis***, which is done by plants. The energy comes from sunlight; photosynthesis cannot proceed without energy from light. Photosynthesis cannot happen in the deep sea for example. The backward direction of the chemical reaction is called ***respiration***; we as consumers do that. We consume the products of photosynthesis (indirectly perhaps if we are not vegetarians), and breathe oxygen, to harvest the energy originally from sunlight, exhaling CO_2 and water vapor.

Photosynthesis serves two purposes in the biosphere. First, it creates carbon in the oxidation state at which its chemistry is versatile enough to build the machinery of life. Second, it stores energy from sunlight in the form of reduced carbon. Some biomolecules have a clear use in the living machine, such as proteins which make enzymes that control chemical reactions, DNA which stores patterns for making proteins, or the fat molecules in cell membranes which hold chemicals inside the cells. Others are more obviously for storage of energy, such as glucose in "blood sugar" or the fat deposits underneath our skin.

Nearly all of the organic carbon produced by photosynthesis is respired sooner or later. Peat deposits on land may hold organic carbon for thousands of years, but on geologic time, the only hope an organic carbon molecule has of escaping degradation back to CO_2 is to hide in the sediments of the deep sea. Life on Earth has built up a sizable pool of carbon in the reduced form in ocean sediments and ***sedimentary rocks*** (former ocean sediments that are currently on land). When oxygen-producing photosynthesis first began on Earth, the oxygen that was released reacted with other reduced species such as dissolved iron in seawater. After the biosphere "rusted" as much as it was going to, oxygen started to build up in the atmosphere. The amount of buried reduced carbon exceeds the amount of O_2 in the atmosphere by about a factor of 10. Oxygen in the atmosphere has built up as a sort of by-product of photosynthesis.

Visionary Earth scientist James Lovelock imagines the biota "charging up" the biosphere like a giant battery, by building up a surplus of reactive oxygen and reduced carbon. Certainly we as oxygen-breathing animals benefit from this situation. Bacteria are able to derive energy by reacting organic carbon with other chemicals, for example, sulfate respiration producing hydrogen sulfide, rotten-egg smelling stuff, to us one of the stinkiest chemicals in nature. Bacteria prefer O_2 for respiration over sulfate or any other chemical because more energy can be gained by using O_2 as one pole of your battery (the food as the other). As for multicellular life, we are unable to respire using anything but O_2.

The land breathes

The CO_2 in the atmosphere is only the tiniest fraction of the carbon on Earth (Fig. 8.1). There is also carbon stored at the land surface, in the oceans, in sedimentary rocks, and the deep interior of the Earth. These other carbon reservoirs all "breathe" CO_2, causing atmospheric CO_2 to vary, naturally, on all sorts of timescales, from one year to millions of years.

The atmosphere contains about 700 gigatons of carbon. A gigaton, abbreviated Gton, is a billion (10^9) metric tons, equivalent to 10^{15} g. The amount of CO_2, relative to the other gases, is about 380 ppm, or 0.038% (percent means "parts per hundred"). For every CO_2 molecule in the air, there are roughly 2500 molecules of other gases, mainly nitrogen and oxygen. If we were able to gather up all of the CO_2 in the atmosphere and

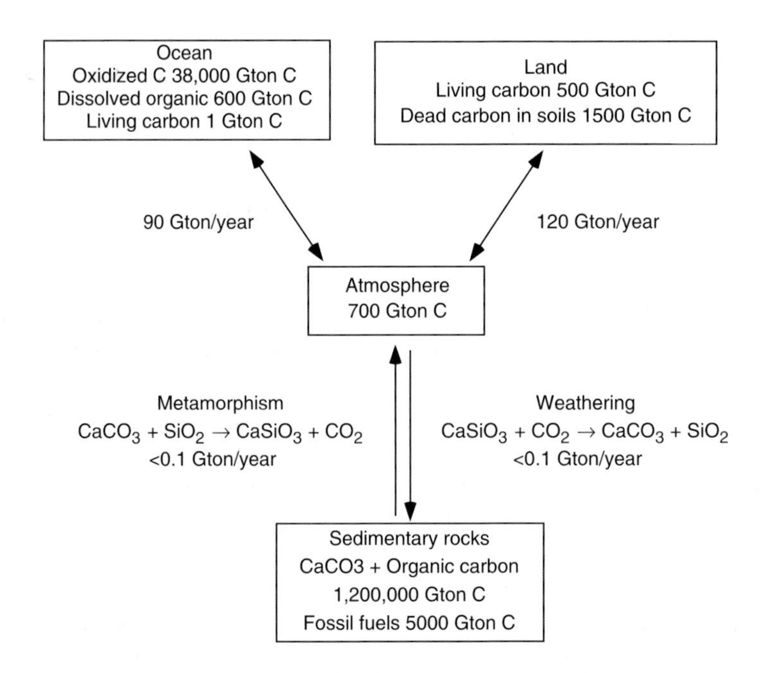

Fig. 8.1 **Carbon reservoirs on Earth and carbon fluxes between them.**

bring it down to the ground to build a column of pure CO_2 gas at sealevel pressure, that column would be about 3 m high, reaching perhaps some second-story windows. If we were to precipitate the CO_2 into a snowfall of dry ice (solid CO_2), we would end up with about 7 cm of snow on the ground. The atmosphere is a very small, very thin carbon reservoir.

There are two forms of carbon that we associate with the landscape that we live in. The actual living carbon, trees and camels and all the rest of it, is called the ***terrestrial biosphere***. There is about 500 Gton C in the terrestrial biosphere, similar in size to the atmosphere. It turns out that there is more carbon in soils than there is in the living biosphere. This is the ***soil carbon pool***: you can see in Fig. 8.1 that it is larger than the terrestrial biosphere carbon pool by about a factor of about two. The soil carbon pool is largely dead carbon, decomposing leaves and other plant material. The amount of carbon stored in soils is highly variable from place to place, depending on the climate, the forestry, and the history of the land. Deserts do not have much carbon in their soils whereas grasslands tend to have quite a bit. There is more soil carbon in colder climates than in warmer climates because organic carbon decomposes more quickly when it is warm. Farming tends to decrease the amount of carbon in soils, but a technique called no-till avoids this problem to some extent.

The terrestrial biosphere inhales CO_2 from the atmosphere during the growing season, and exhales it during winter. You can see a seasonal cycle in atmospheric CO_2 in Fig. 8.2. The seasons in the southern hemisphere are the reverse of those in the northern hemisphere, and you can see that the CO_2 cycles are reversed also. The cycles are smaller in the southern hemisphere than they are in the north because there is less land in the south. One interesting observation, foreshadowing Chapter 10, is that the annual cycles in atmospheric CO_2 have been getting larger as atmospheric CO_2 has risen. It's not obvious by eyeballing Fig. 8.2, but the trend emerges when the data are treated statistically. The terrestrial biosphere has been breathing more deeply with increasing atmospheric CO_2.

The ocean breathes

The ***ocean carbon reservoir*** is larger than either the land surface or the atmosphere. The ocean's carbon is not only dead, but it is also oxidized: energetically dead as well as biologically dead. The carbon is in the forms CO_2, H_2CO_3, HCO_3^-, and $CO_3^=$, all of which we will get to know better in Chapter 10. The sum of all of them is called ***dissolved inorganic carbon***. The ocean dissolved inorganic carbon pool is larger than the atmospheric pool by a factor of about 50. There is also dissolved organic carbon in the ocean: dead, scrambled molecules similar to soil organic carbon. And of course there are fish and dolphins and plankton, but the total amount of carbon in the ocean that lives is small, only about a Gton.

Carbon is released from the water to the air in some parts of the world ocean and dissolves into the water in other places. The total rates of exchange are large, comparable with the breathing rates of the terrestrial biosphere and much greater than the rate of anthropogenic carbon release (Chapter 9). The oceans would seem to have the

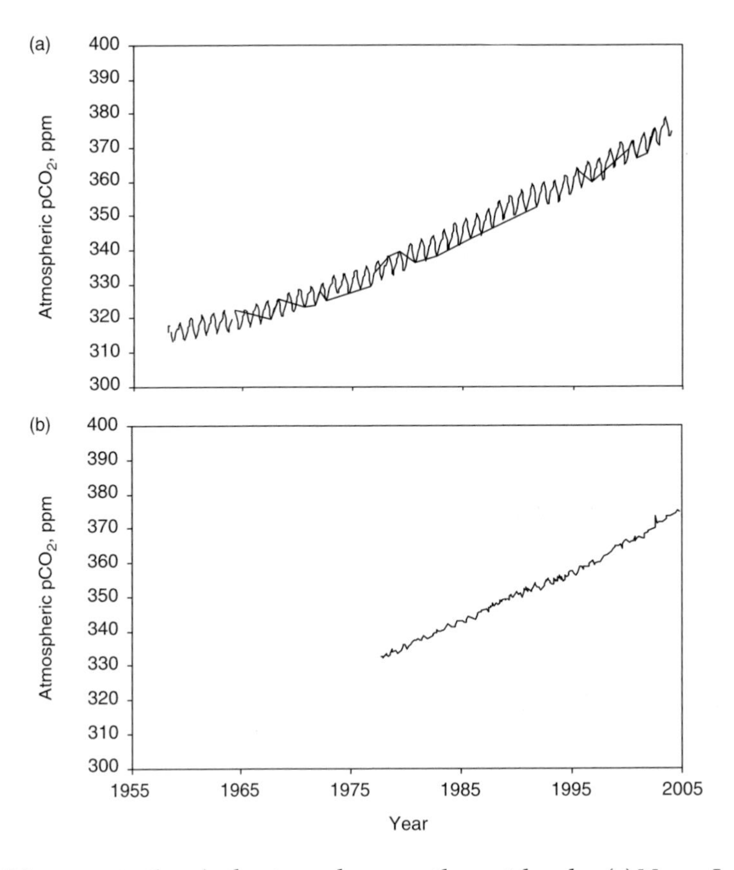

Fig. 8.2 **CO_2 concentrations in the atmosphere over the past decades. (a) Mauna Loa, Hawaii and (b) Baring Head, New Zealand.**

clout to affect the CO_2 in the atmosphere as strongly as the terrestrial biosphere does, and yet the year-to-year breathing variations in the ocean are much less obvious in variations in atmospheric CO_2. The ocean simply breathes more slowly than the land does.

The atmosphere absorbs CO_2 from the ocean until the rates of CO_2 into and out of the ocean balance each other. Computer model experiments show that it takes hundreds of years for the atmospheric CO_2 concentration to approach this equilibrium value. The reason it takes so long is that the ocean takes about this long to circulate, for all the waters of the ocean to come to the sea surface somewhere to exchange carbon with the atmosphere.

The clearest example of the power of the ocean to affect CO_2 in the atmosphere is the ***glacial/interglacial cycles***. The geologic record of the last 500 million years shows Earth going into an ***ice age*** every 150 million years or so. We are in an ice age now, and have been for about 2 million years. During an ice age, the amount of ice, and the climate of the Earth, fluctuate rhythmically between glacial states and interglacial states (Fig. 8.3). We are currently in an interglacial state, having been for 10,000 years. During the ***Last Glacial Maximum***, global temperature was 5–6 K colder than today

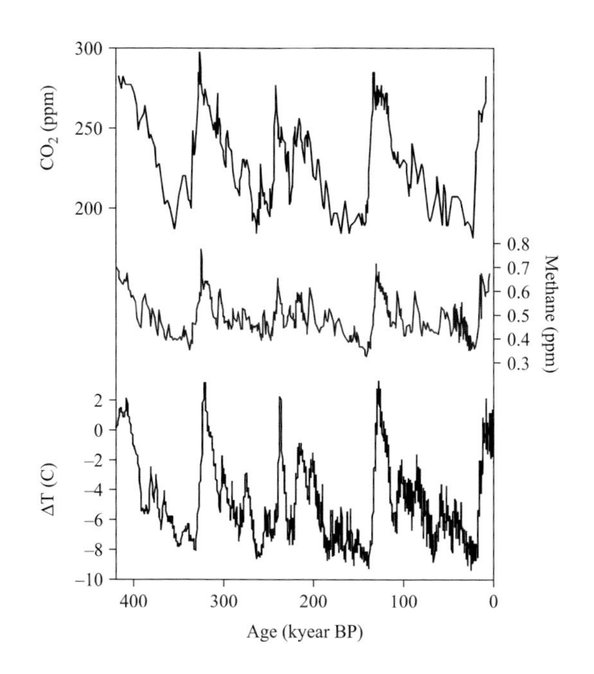

Fig. 8.3 **CO_2 and methane concentrations in the atmosphere over the past 400,000 years, along with temperature in Antarctica.**

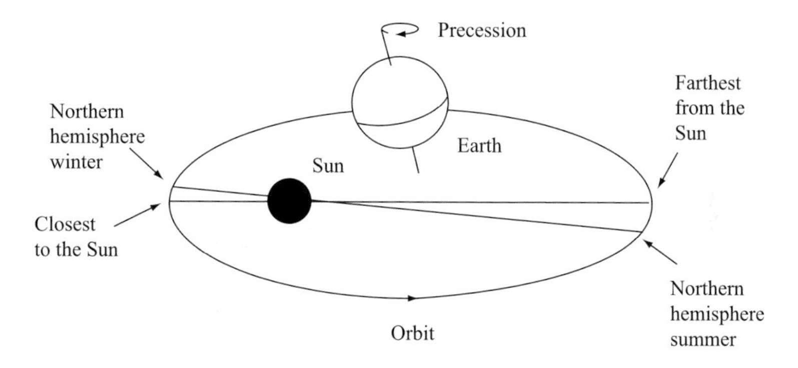

Fig. 8.4 **The precession orbital cycle.**

and much of North America and Northern Europe were covered with a massive ice dome, like what currently exists in Greenland.

The beat of the ice age rhythm apparently derives from variations in the Earth's orbit around the Sun. The orbit varies through three main cycles, each ringing on its own characteristic frequency. The first cycle is called the ***precession*** of the seasons or sometimes precession of the equinoxes (Fig. 8.4). The axis of rotation of the Earth spins around like a wobbling top, completing the entire circle in 20,000 years. Most of the solar heat influx variability at high latitudes derives from precession, and nearly all of the variability in the tropics comes from precession.

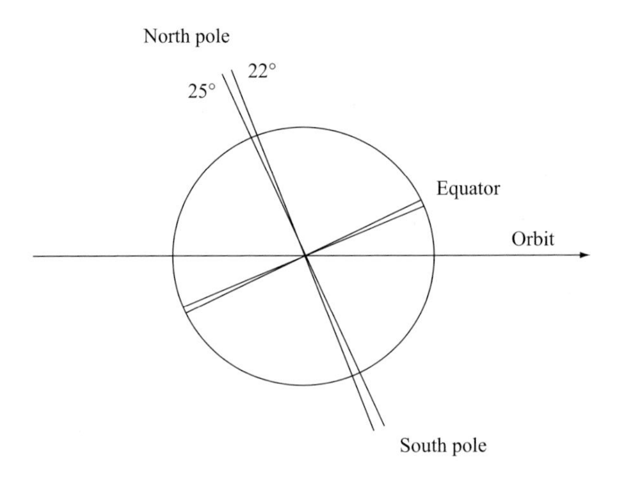

Fig. 8.5 **The obliquity orbital cycle.**

Precession has an effect on climate because the Earth's orbit is not circular but elliptical. Where we are in the precession cycle at present, the Earth is closest to the Sun during winter in the northern hemisphere (Fig. 8.4). The seasonal cycle of solar heat flux in the northern hemisphere is weakened by this orientation, because the Earth is close to the Sun when the northern hemisphere is tilted away from the Sun. The seasonal cycle of solar heating is stronger in the southern hemisphere now because the Earth is close to the Sun and tilted toward the Sun at the same time. Remember, it is the tilt of the Earth that causes the Earth's seasons (Chapter 6). The precession cycle merely modifies the seasons somewhat. The history of precession cycle over the past 400,000 years is shown in Fig. 8.6.

Another cycle involves the ***obliquity*** of the angle of the pole of rotation, relative to the plane of Earth's orbit (Fig. 8.5). The Earth rotates, making day and night, on a rotation axis of the north and south poles. This rotational axis is not perpendicular to the plane of Earth's orbit around the Sun, but is tilted somewhat. The angle of tilt is currently 23.5°, but it varies between 22° and about 25.5°, on a cycle time of about 41,000 years (Fig. 8.6). The impact of obliquity on the solar heating flux is stronger in high latitudes.

The third orbital cycle involves how elliptical the orbit of the Earth is, also called its ***eccentricity***. At present, the orbit of the Earth is nearly circular. The eccentricity of the orbit has cycles of 100,000 and 400,000 years (Fig. 8.6). The strongest climate impact of eccentricity is to determine the strength of the precessional forcing. If the Earth's orbit were circular (as it is nearly now), it would make no difference where the Earth was in its precession cycle because the Earth is equally far from the Sun at all parts of the orbit. When eccentricity is low, the orbit is circular, and the 20,000 year waves in the precession cycle vanish (shaded area in Fig. 8.6).

When you average over the entire surface of the Earth and over the entire year, the orbital cycles only have a tiny effect on the amount of heat the Earth gets from the Sun. The orbital cycles affect climate by rearranging the intensity of sunlight from one place to another, and from one season to another. It turns out that the climate of the Earth is especially sensitive to the solar heat flux at about 65° latitude in the northern hemisphere summer. This is like the solar plexus of global climate; a sucker

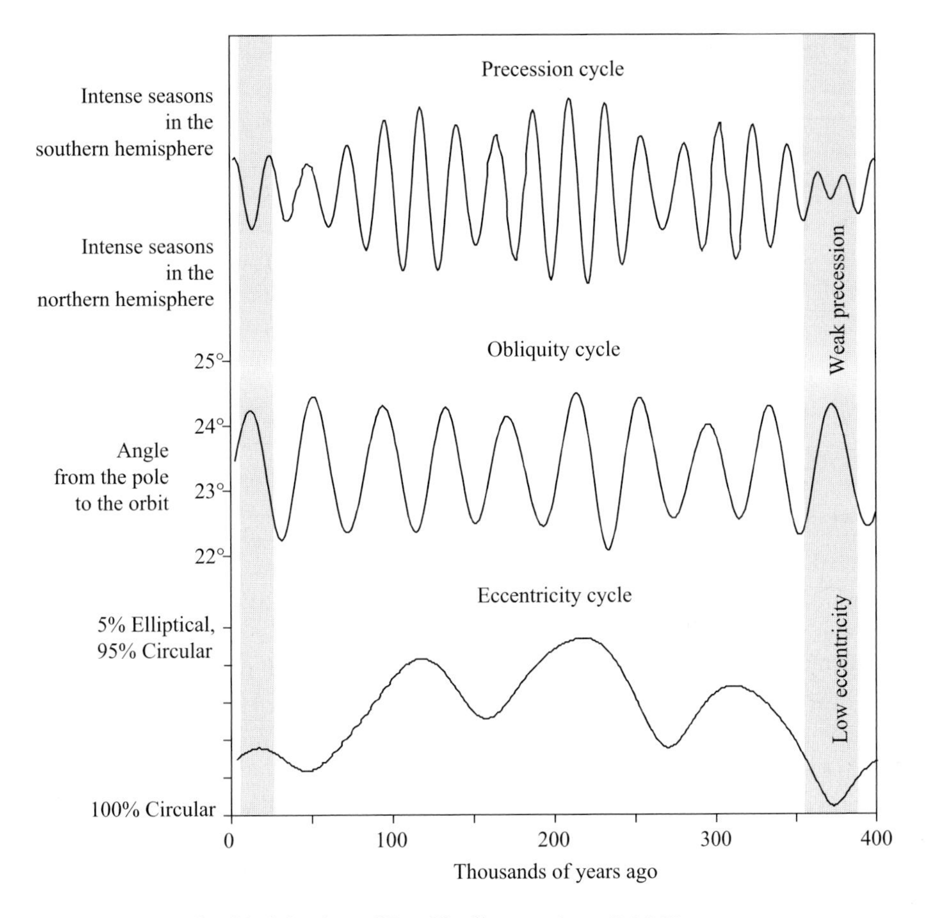

Fig. 8.6 **History of orbital forcing of Earth's climate, since 400,000 years ago.**

punch there and the whole climate keels over. This is because the northern hemisphere is where the ice sheets are that come and go with glacial climate. The ice sheets in Antarctica and Greenland persisted through recent interglacial climate stages, although they do change their sizes. The production of an ice sheet in the northern high latitudes drags the whole world into a glacial climate. The intensity of summertime sunlight seems to be important, rather than wintertime, because it is always cold enough to snow in winter at these latitudes. The question is whether the summer is warm enough to melt the snow in the summer. Variations in sunlight intensity in June, 65° north latitude, calculated from models of the Earth's orbit, correlate well with the history of the amount of ice on Earth, inferred from ice core and deep sea sediment records.

The link from glacial cycles to the CO_2 in the atmosphere and the ocean comes from bubbles of ancient atmosphere trapped in the ice sheet of Antarctica. The trapped air bubbles from glacial times have a lower proportion of CO_2 in them, relative to the other gases in the bubble. The CO_2 concentration during glacial intervals is 180–200 ppm, rising to 260–280 ppm during interglacial intervals before the industrial era. The decreased CO_2 concentration during glacial time is responsible for about half of the cooling relative to the interglacial time. The increase in the albedo of the Earth, from the large ice sheets, is responsible for the other half. The climate during glacial time is an important test of the climate models we use to forecast global warming (Chapter 11).

No one is sure exactly why CO_2 in the atmosphere cycles up and down along with the ice sheets, but the ocean must be the major player. The land carbon reservoirs were if anything smaller during the glacial time, which by itself would have left atmospheric CO_2 higher, not lower as we see. The carbon in rocks, discussed in the next section, cycles too slowly to explain the large fast changes. The ocean is the only pool that is large enough and potentially reactive enough to explain the data. We will not answer the question of how exactly the ocean did this, but we will discuss ocean carbon chemistry a bit further in Chapter 10.

The rocks breathe

The ***sedimentary rock carbon pool*** is larger still than the ocean, land, or atmospheric pools. Carbon exists in the form of ***limestones***, $CaCO_3$, and to a lesser extent as organic carbon. These carbon reservoirs together contain about 500 times as much carbon as the atmosphere and the landscape combined. Most of the organic carbon in sedimentary rocks is in a form called ***kerogen***. Kerogen is useless as a fossil fuel because it is dilute, usually less than 1% by weight of sedimentary rocks, and because it is in a solid form making it difficult to extract. We will come back to fossil fuel forms of carbon in Chapter 9.

Carbon is exchanged between the atmosphere and the sedimentary $CaCO_3$ rocks by means of a chemical reaction called the ***Urey reaction***, which is

$$CaSiO_3 + CO_2 \leftrightarrow CaCO_3 + SiO_2$$

and as shown in Fig. 8.7. $CaSiO_3$ on the left-hand side is a simplified chemical formula for a rock formed at high temperature called a ***silicate rock***. Silicate rocks are formed by cooling and freezing melted rock. Melted rock is called ***lava*** if it is found at the Earth's surface and ***magma*** if it is in the subsurface. Real silicate rocks have other elements in them and a wide range of chemical formulas, but the simple formula $CaSiO_3$ works for conveying the essential idea. $CaCO_3$ and SiO_2, the solid phases on the right-hand side of the reaction, are typical sedimentary rocks, which form at cold temperatures from elements that were dissolved in water.

The Urey reaction running from left to right, producing sedimentary rocks from silicate rocks, is called ***weathering***. A silicate rock weathers by dissolving its calcium and silica into river water, ultimately to be delivered to the ocean. Left to right is the preferred direction for this chemical reaction, under the relatively cold, wet conditions of the surface of the Earth. If weathering were allowed to continue to its equilibrium, it would pull nearly all of the CO_2 out of the atmosphere. Organisms like corals and shell-forming plankton extract the dissolved ions from seawater and construct solid $CaCO_3$ and SiO_2 from them.

The Urey reaction runs as we have written it backward, from right to left, producing silicate rocks from sedimentary rocks, deep in the Earth's interior where it is hot. This

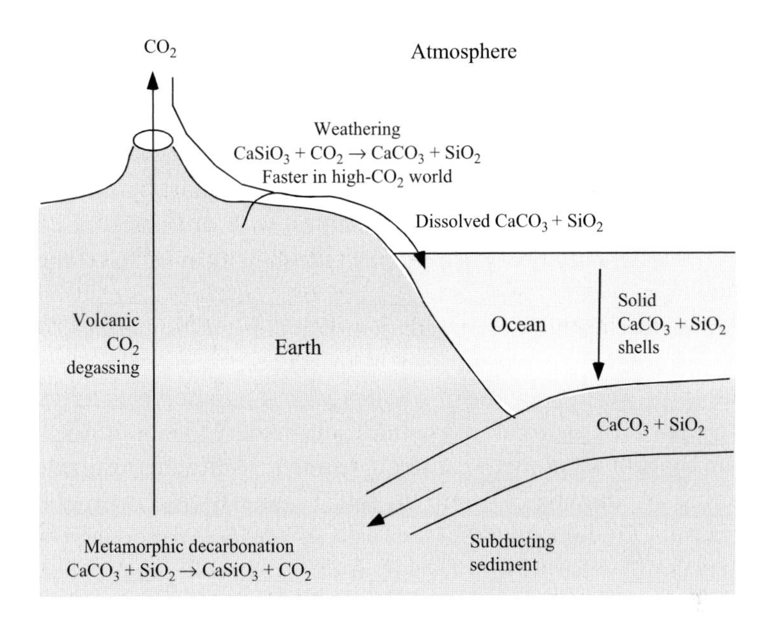

Fig. 8.7 **Components of the silicate weathering thermostat.**

chemical reaction is called ***metamorphic decarbonation***. This direction is generally favored at high temperature. CO_2 released by metamorphic reactions may find its way to the surface in volcanic gases or in hot water springs at the bottom of the ocean. Some of the carbon degassing from the Earth may be ***juvenile*** carbon, which has spent the last 4.5 billion years of the Earth history bound up in the deep Earth, only to emerge now. We will refer to CO_2 fluxes from the Earth, juvenile and metamorphic, as ***volcanic CO_2 degassing***.

The fluxes of CO_2 by weathering and degassing are small compared with the other fluxes in Fig. 8.2, but if they were to be out of balance, that is, if you were to stop all degassing for example, you could use up all the CO_2 in the atmosphere in a few hundred thousand years. The Earth is much older than this, so if we average over a million years or longer, the weathering and degassing fluxes of CO_2 must balance. CO_2 that escapes from the Earth must eventually be consumed by weathering, since what comes in must go out.

The way the Earth manages to balance the degassing and weathering fluxes is by finding the CO_2 concentration and the climate at which the rate of weathering balances degassing. The rate of weathering depends on the availability of fresh water, from rainfall and runoff, that rocks can dissolve into. The rate of fresh water runoff depends, in turn, upon the climate of the Earth. If the climate gets too cold, then weathering slows down, allowing CO_2 in the atmosphere to build up. If the climate is too warm, CO_2 will be consumed by weathering faster than it is degassed from the Earth.

This need to balance the degassing and weathering CO_2 fluxes acts to stabilize the CO_2 concentration of the atmosphere and the climate of the Earth. This climate-stabilizing mechanism is called the ***silicate weathering thermostat***. We can recycle the sink analogy from Chapter 3, only now the faucet is CO_2 degassing, the drain is silicate weathering, and the water level in the sink is CO_2 in the atmosphere. If CO_2 is too high, it will be drawn down by weathering, or if it is too low, it will accumulate from volcanic degassing, until the fluxes of CO_2 balance.

The catch to this mechanism is that it takes hundreds of thousands to years to stabilize the CO_2 concentration and climate at their equilibrium values. On timescales shorter than that, it is perfectly possible for perturbations, such as the glacial cycles or the fossil fuel CO_2 release. This theory helps explain the stability of Earth's climate through geologic time.

The silicate weathering thermostat leads to variations in atmospheric CO_2 concentration, on timescales of millions of years. CO_2 changes occur because other factors besides CO_2 may affect the weathering rate of silicate rocks. A mountainous terrain weathers more quickly than a flat plain covered in thick soil, because the soil isolates the silicate bedrock from the rain water that weathering requires. Plants affect the rate of silicate rock weathering by pumping CO_2 into the soil. There may also be variation through time in the rate of CO_2 degassing from the Earth, driven by variation in plate tectonics.

Some intervals of the Earth's history, such as the ***Cretaceous period*** when dinosaurs ruled the Earth, and ***Early Eocene optimum*** which followed, apparently needed more CO_2 in their atmosphere in order to balance the degassing and weathering CO_2 fluxes at that time. The breathing of the rocks, via the silicate weathering thermostats, takes place on timescales of millions of years (Fig. 8.8).

The terrestrial planets Venus, Earth, and Mars provide a perfect Goldilocks parable to end this section. The sin of Venus (the planet) was to be too close to the Sun. The water that must have originally been packed with Venus evaporated and its hydrogen lost to space forever, a runaway greenhouse effect discussed in Chapter 7. With no water, silicate weathering reactions do not go. With weathering out of the picture, degassing

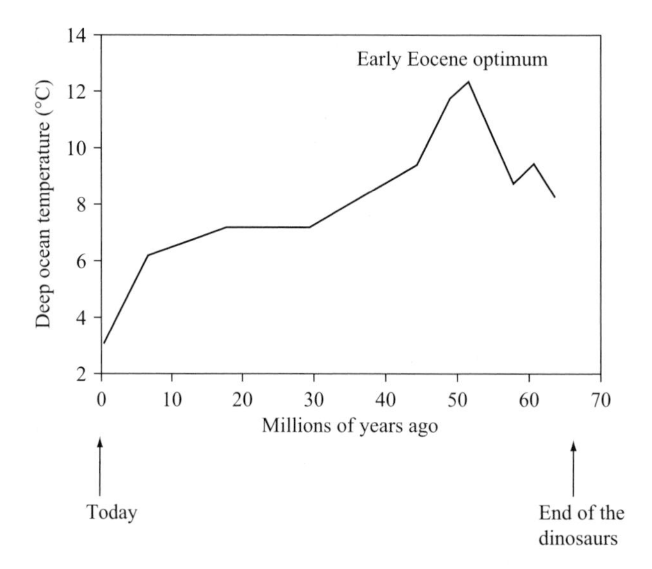

Fig. 8.8 **History of the temperature of the deep ocean, which tells us about the temperature of the high latitude surface, since 65 million years ago.**

wins. Venus' carbon allotment ended up as CO_2 in the atmosphere, 70 atm of CO_2. At that pressure, CO_2 is no longer strictly speaking a gas; the atmosphere of Venus is actually more like an ocean. And of course Venus is very hot.

Mars messed up by being small, about half the diameter of Earth, so that its interior cooled faster than Earth's. Mars boasts the largest volcano in the solar system, Olympus Mons, but today Mars is geologically dead. With degassing out of the picture, weathering wins and the carbon allocated to Mars has ended up as $CaCO_3$ rocks. And Mars is cold.

Take-home points

1. The most stable form of carbon on Earth is oxidized, as CO_2 or $CaCO_3$. Photosynthesis stores energy from the Sun by producing organic carbon, which also serves as a building scaffolding for the machinery of life.
2. There is less carbon in the atmosphere than there is in other carbon reservoirs on Earth, such as the terrestrial biosphere and the oceans. These other reservoirs tug on atmospheric CO_2, seasonally for the land, and on 100,000-year glacial/interglacial cycles from the ocean.
3. The weathering of igneous rocks on land controls the pCO_2 of the atmosphere on million year timescales. This "silicate thermostat" stabilizes climate. The thermostat is broken on Venus, because there is no water left, and on Mars, because there is no active volcanism left.

Projects

Answer these questions using an online geologic carbon cycle model located at http://understandingtheforecast.org/Projects/geocarb.html. The model predicts the influence of geologic processes on the CO_2 concentration of the atmosphere. To begin with, the atmosphere and the ocean exchange CO_2. It takes about 300 years for CO_2 to find equilibrium between the ocean and the atmosphere, if the chemistry of the ocean was held constant. Chemical reactions with rocks on land and on the sea floor further alter the carbon chemistry of the atmosphere and ocean on much longer timescales. One set of chemical reactions involves $CaCO_3$, which must be removed from the ocean by burial ("burial of carbonates") as quickly as it is supplied to the ocean by dissolution of rocks on land ("total weathering"). It takes roughly 5–10 kyear for the ocean to balance its $CaCO_3$ budget, to come to $CaCO_3$ equilibrium. The $CaCO_3$ equilibrium adjustment changes the pH of the ocean and the pCO_2 of the atmosphere. The other geochemical cycle affecting CO_2 involves igneous rocks. The silicate thermostat acts to balance the weathering of silicate rocks against degassing of CO_2 from the Earth, via volcanos and hot springs.

The model does an initial spinup phase to make sure that the initial condition is an equilibrium where all the various budgets balance. You will only see the last few years of this phase in the model output. Then there is a transition to a new set of conditions, say an increase in CO_2 degassing flux from volcanos. We can watch the transition from one equilibrium state to another.

At the moment of the transition from spinup phase to transient phase, we can also inject a slug of new CO_2 into the system, say by fossil carbon combustion and release to the atmosphere. The high CO_2 concentration at the top of the spike depends a lot on how quickly the CO_2 is released, which is unrealistically quickly in this model, so don't take the spike magnitudes to the bank. Pay more attention to what happens to CO_2 in the long term after it is released.

1. *Weathering as a function of CO_2.* The rate of weathering must balance the rate of CO_2 degassing. Run a simulation where the CO_2 degassing rate increases or decreases at the transition time. Turn off the CO_2 spike (set it to zero), to make things simpler. Does an increase in CO_2 degassing drive atmospheric CO_2 up or down? Repeat this run with a range of final degassing rates, and make a table of the CO_2 concentration as a function of the CO_2 degassing rate. The CO_2 degassing rate is supposed to balance the CO_2 consumption rate by silicate weathering – verify that the model achieves this. If so, make a plot of weathering as a function of atmospheric pCO_2.
2. *Effect of solar intensity.* The rate of weathering is a function of CO_2 and sunlight, a positive function of both variables. By this I mean that an increase in CO_2 will drive an increase in weathering, as will an increase in sunlight. The Sun used to be less intense than it is now. Turn back the clock 100 or 500 million years, to dial down the Sun. Weathering has to balance CO_2 degassing, so if sunlight goes down, CO_2 must go up, to balance. Try it. What do you get for the initial steady-state CO_2, relative to what you get for today's equilibrium value?
3. *Plants.* Plants pump CO_2 down into the soil gas, possibly accelerating weathering. They also stabilize soils, perhaps decreasing weathering. Run a simulation with a transition from no plants to a world with plants, with no carbon spike on the transition. Figure out whether plants in the model, overall, increase or decrease weathering.

Further reading

Alley, Richard B., *The Two-Mile Time Machine: Ice Cores, Abrupt Climate Change, and Our Future*, Princeton, 2002.

9
Fossil fuels and energy

The Sun bathes the Earth in an immense amount of energy, but most of humankind's energy sources derive from stored energy sources such as fossil fuels or radioactive elements rather than instantaneous solar energy. Of the fossil fuels, coal is the most abundant whereas oil is more limited and may be depleted in the coming decades. The amount of natural gas as traditionally extracted is comparable with that of oil, but there is an immense amount of gas frozen in ocean sediments. If we wish to stabilize the CO_2 concentration of the atmosphere in the coming century, we need some major new carbon-free source of energy, of a size comparable to our total energy production today or larger.

Energy sources

Our civilization is based on energy. We will express global energy fluxes in units of ***terawatts***. A watt is a flux of energy flow, equal to joule per second, where joule is a unit of energy like calories. A hair dryer might use 1000 W of power, or 1 kilowatt, kW. A terawatt is 10^{12} W, and is abbreviated as TW.

The Sun bathes the Earth in energy at a rate of about 173,000 TW. Photosynthesis captures about 100 TW. Mankind is consuming about 13 TW of energy per year. Some of our 13 TW of energy use comes out of the 173,000 TW the Earth gets from the Sun. Windmills, hydroelectric dams, solar cells, and firewood all derive their energy relatively directly or indirectly from sunlight. But this is not enough; we are also taking advantage of other forms of stored energy, from fossil fuels and radioactive elements. Fossil fuels derive their energy from past sunlight, photosynthesis that took place millions of years ago.

Over the Earth's history, photosynthesis has preserved a sizable amount of reduced carbon, which could in principle be reacted with O_2 to regenerate CO_2 and energy. If we tried it on Earth, we would run out of oxygen in the air before we ran out of reduced carbon. As it turns out the vast majority of this reduced carbon is so diluted by rocks and clay minerals that it is unusable for fossil fuels. Only in special conditions do we find organic carbon concentrated enough to be useable for energy. The main forms of fossil carbon that we use for converting to energy are coal, oil, and natural gas.

Globally, most of the energy we use comes from fossil fuels such as petroleum, gas, and coal (Fig. 9.1). Methane is the most reduced form of carbon, so that when an atom of carbon from methane undergoes the transition to oxidized CO_2, it gives off more energy than would an atom of carbon in coal, which has an oxidation state at the level

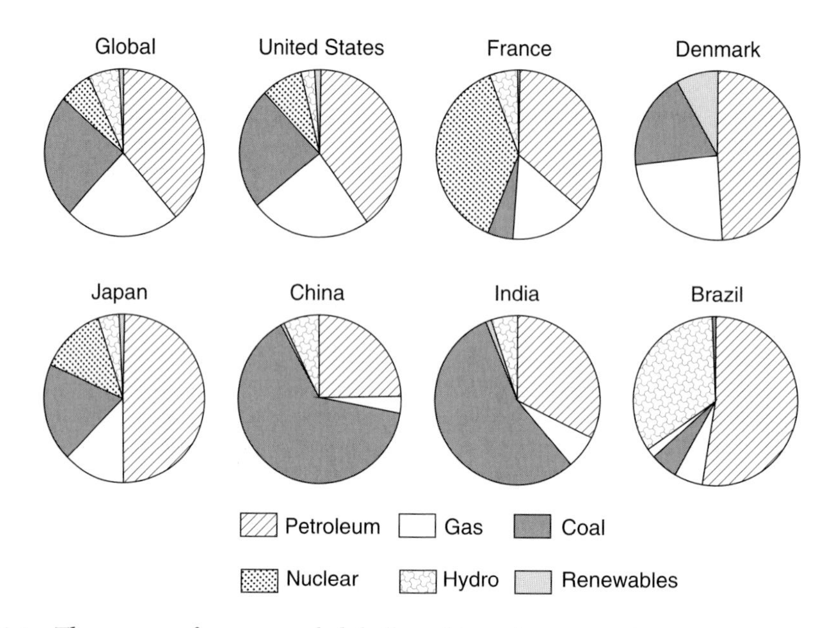

Fig. 9.1 **The sources of energy used globally and in various countries.**

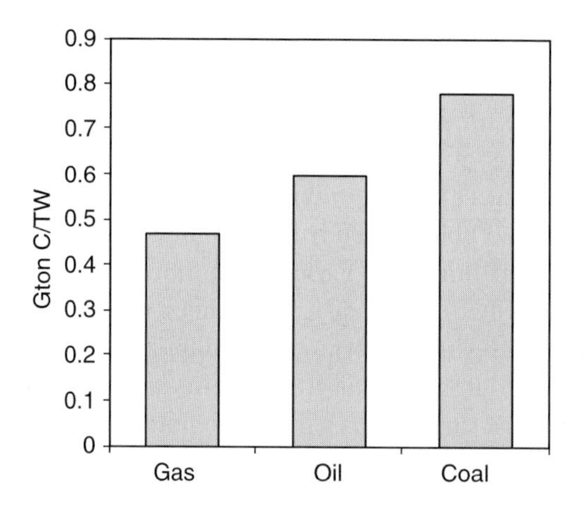

Fig. 9.2 **Natural gas emits less CO_2 per watt of energy than oil or coal.**

of carbohydrates (O). This means that a watt of energy from methane releases less carbon than a watt from coal or wood (Fig. 9.2).

Non-carbon-emitting sources, mostly nuclear and hydroelectric, make up about an eighth of the global energy supply. We use renewable sources of energy, defined by the US Department of Energy to mean geothermal, solar, wind, and wood and waste electric power, for about 0.8% of our energy.

The largest traditional fossil fuel reservoir is ***coal***. Coal is solid, almost pure carbon when it is mature, the good stuff. Coal originates from land plants that deposited in swamps. Land plant deposits in swamps today, that might turn into coal someday, are

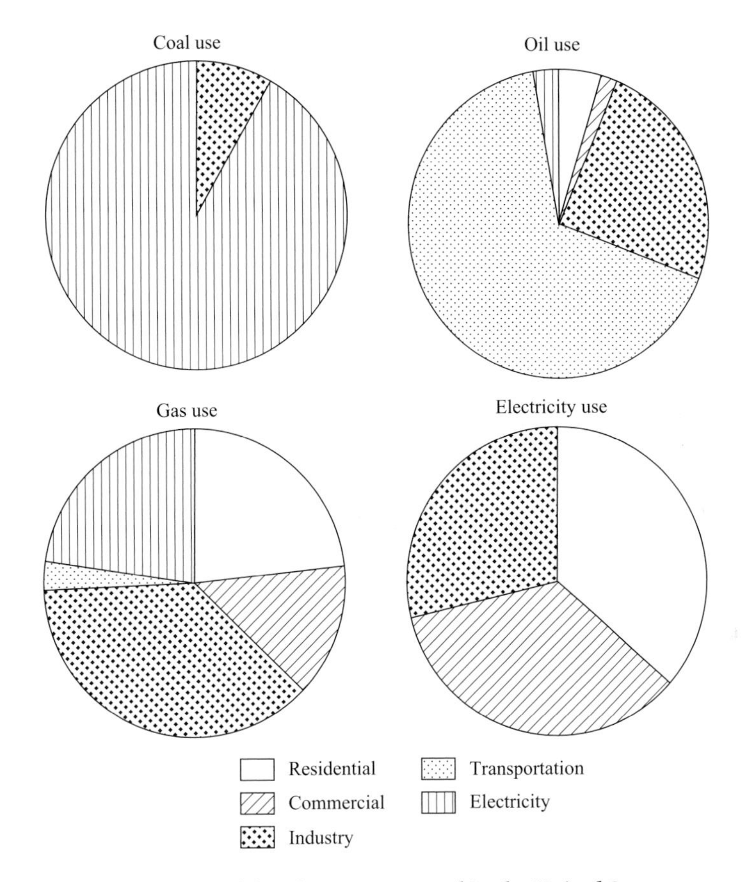

Fig. 9.3 **How different forms of fossil energy are used in the United States.**

called ***peat***. The British Petroleum (BP) company, in their annual report on global energy (see Further reading) estimates the existence of 1000 Gton of carbon in coal reserves. Another source, an academic paper by Rogner (also listed in Further reading), estimates that the total amount ultimately available might be 10 times that. No matter what assumptions are made about future energy needs, there is enough coal to supply our needs for several centuries, if we choose to use it. In the United States, coal is mostly used for generating electricity (Fig. 9.3).

Oil and ***natural gas*** mostly come from ancient photosynthesis that took place in the ocean. Plants in water are generically called ***algae***. There are microscopic algae in the ocean called ***phytoplankton***. The phytoplankton in the ocean produce about as much organic carbon by photosynthesis as the terrestrial biosphere on land does. A tiny fraction of the dead phytoplankton ends up in sediments, while the rest gets eaten by somebody, animal or bacterial, in the water column or in the top layer of sediments. The sediments covering most of the sea floor do not contain enough organic carbon to ever make oil. You find organic-rich sediments nearby continents and under waters which have no oxygen dissolved in them, that is, in ***anaerobic*** conditions.

If organic-rich sediments are buried to a depth in the Earth between 7 and 15 km, they will be heated to temperatures of 500–700 K. This has the effect of converting some

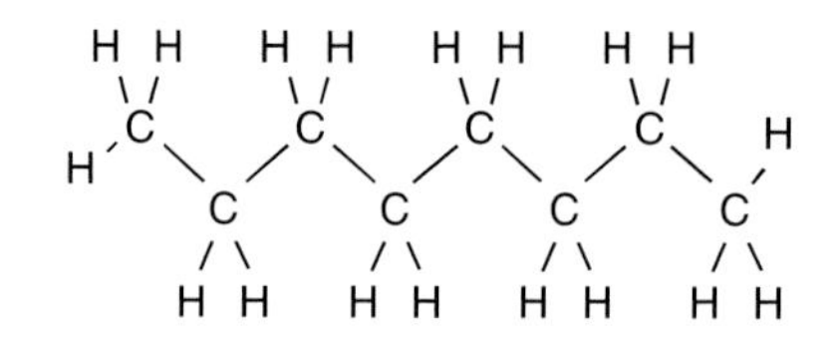

Fig. 9.4 **The structure of normal octane, a hydrocarbon chain such as found in petroleum.**

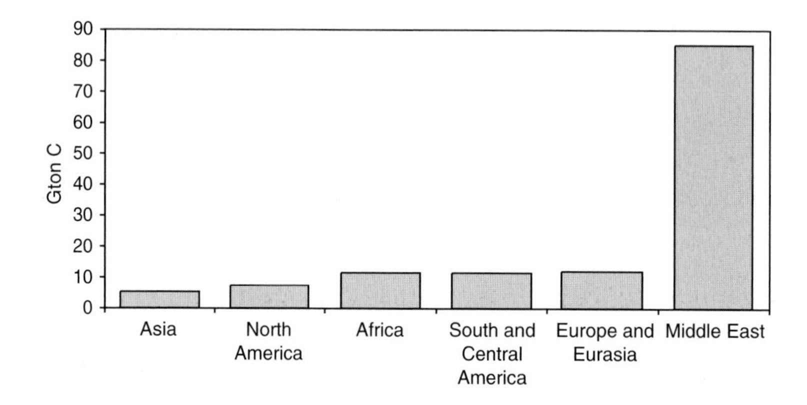

Fig. 9.5 **The amount of proven petroleum reserves, by region, from BP.**

of the dead plankton into oil, consisting of chains of carbons with hydrogens attached (Fig. 9.4). The oil may just sit there in little droplets, unreachable by oil companies, or it may flow through the rock if it is porous enough, perhaps to collect in some upside-down pool on the bottom of an upside-down bowl in some impermeable layer of rocks above it. Probably only a tiny fraction of the oil ever produced is in a harvestable form today, that is to say mobile enough to have flowed together but not mobile enough to get squeezed all the way out of the Earth. Oil is perhaps the most convenient of the fossil fuels because it is easily stored and transported in its liquid form. In the United States, oil is mostly used for transportation (Fig. 9.3).

Perhaps because such special conditions are required to produce harvestable oil, the distribution of oil deposits on Earth is very spotty (Fig. 9.5). Most of the oil is in the Middle East, although there is a smattering scattered throughout the rest of the world. We don't know precisely how much oil there is available to harvest. The BP report tallies the proven reserves of oil, stuff that has been discovered and documented already (Fig. 9.5). They say 1,150,000 million barrels of oil, which we can translate into 135 Gton C. The amount of oil that is likely to be ultimately extractable is almost certainly higher than this. Technological advances in oil extraction can double the amount of oil you can get from an oil field. More oil fields will certainly be discovered, although there are limits to how much more oil we can expect to discover. Half of the world's oil reserves today are in the 100 biggest oil fields, and most of the largest ones were discovered decades ago.

There is a class of carbon deposits for which the geological conditions were never quite right to produce oil, but from which oil can be extracted if the rocks are mined and

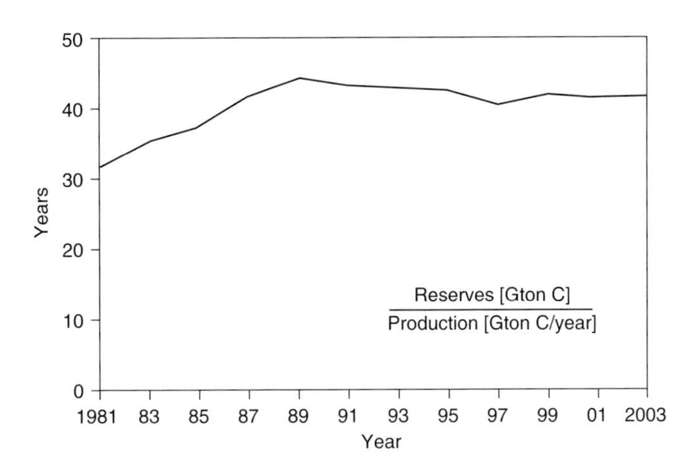

Fig. 9.6 **The value of the production to reserves ratio over the last 25 years, from BP.**

cooked. These nontraditional oil sources are called ***oil shales*** and ***oil sands***. The German government during the Second World War tried desperately to produce oil from coal but never succeeded. The Canadians are mining and processing oil shales now. If future discoveries, technology progress, and all these effects are counted together, the estimates are all over the map, but some of them begin to reach 500 Gton C or so.

How long will the oil last? This is a question that is starting to be discussed in the newspapers. One way to calculate a number with units of time would be

$$\text{Lifetime[year]} = \text{reserves size[kg]} \cdot \frac{1}{\text{production rate}} \left[\frac{\text{year}}{\text{kg}} \right]$$

BP calls this the reserves to production ratio. The R/P ratio tells us how long it would take to use up a resource if we continued using it at the rate we're using it now. Notice that the units on both sides of the equation balance. The funny thing about this ratio is that its value hasn't changed in 15 years (Fig. 9.6). According to BP figures, we have 40 years of oil left, and it looks like we always had 40 years left, and we could have 40 years left forever. The clock seems to be ticking very slowly. Actually, what is happening is that oil is being discovered, and extraction efficiency is increasing, about as quickly as oil is being used. The rate of consumption is growing with time, but we have been discovering new oil quickly enough to keep pace with that growing demand.

Another way to estimate the lifetime of the age of oil was developed by geologist M. King Hubbert (see the Deffeyes books listed in Further reading). Hubbert pointed out that the rate of extraction of a limited natural resource such as oil tends to follow a bell-shaped curve. This is an example of an ***empirical*** observation; there is no theoretical underpinning to the curve, but it sure does fit nice. The rate of extraction has a spinup time at the beginning, gets going faster and faster, then slows down as the resource starts to run out. Resource extraction histories are not required to follow this curve. One could imagine an extract-the-juice-from-a-popsicle-on-a-hot-day curve. The popsicle consumption rate starts off slowly at the beginning because the

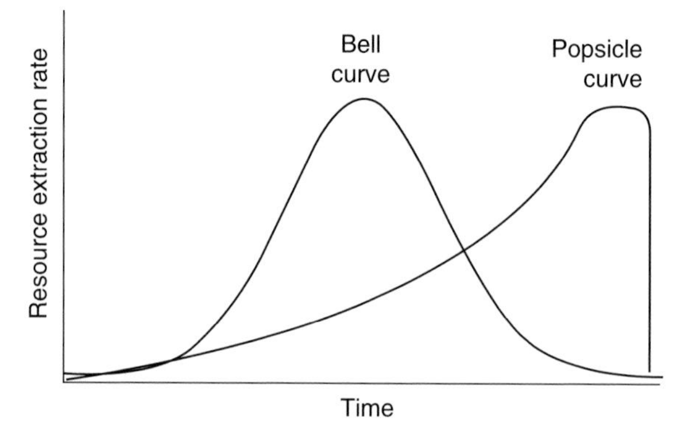

Fig. 9.7 **A bell curve and an eating-a-popsicle-on-a-hot-day curve.**

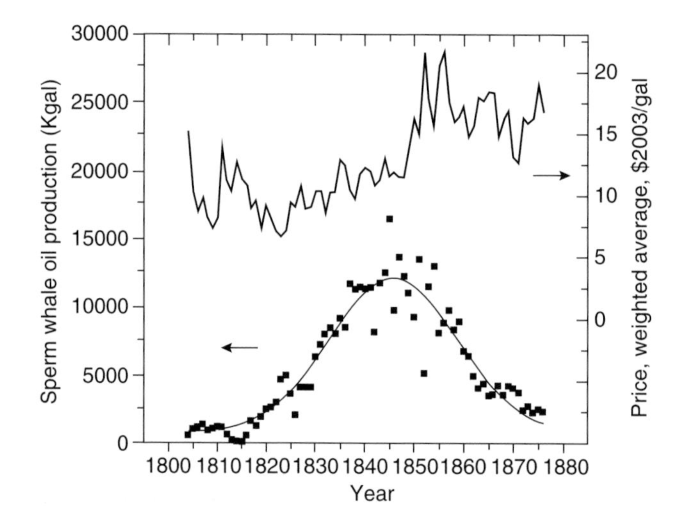

Fig. 9.8 **A Hubbert curve for the rate of sperm whale oil production. With permission from U. Bardi, Università di Firenze.**

popsicle is too cold to eat, then at the very end you have to swallow the last half of the popsicle in one gulp to keep it from hitting the sidewalk (Fig. 9.7). The crucial feature of the bell curve is that the maximum rate of extraction, the peak, occurs when half of the resource is still in the ground. The peak of the popsicle-eating curve comes at the very end. It could be that global oil will follow a popsicle curve instead of a bell curve, but a bell curve seems to have worked in the past, for individual oil and coal fields. Figure 9.8 shows a Hubbert's peak for sperm oil harvesting back in the whaling days.

Hubbert's theory gains credibility from the fact that Hubbert used it, in 1956, to forecast the peak extraction of oil in the United States. He predicted the peak to be sometime between 1965 and 1972, and in fact it came in 1970. The solid line in Fig. 9.9 shows the time up to Hubbert's forecast, and the dashed line a prediction like his for the future (benefiting a bit from hindsight, I confess). Hubbert's prediction came at a time when oil production in the United States seemed endless.

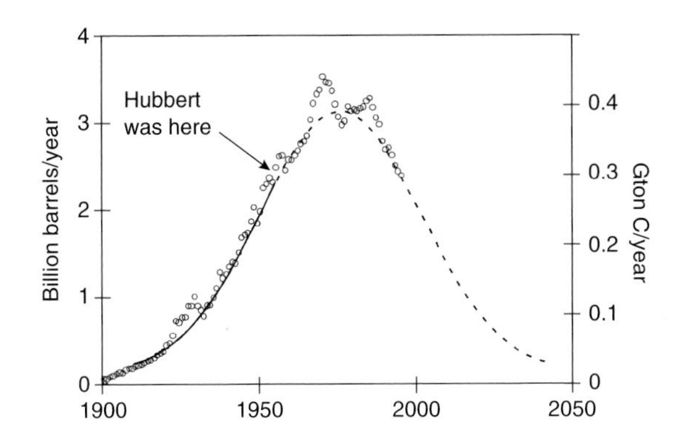

Fig. 9.9 **Oil production in the United States follows a bell curve.**

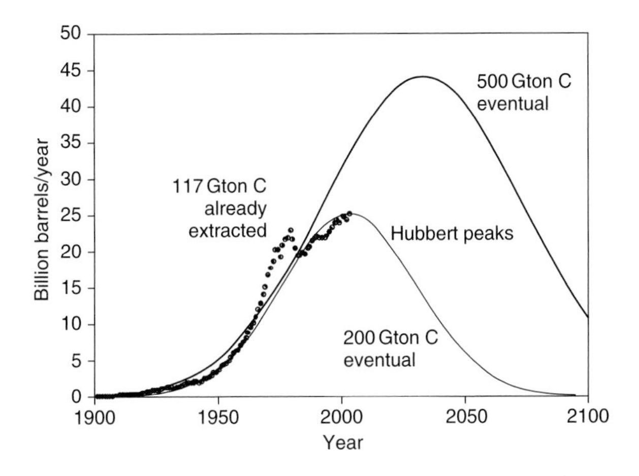

Fig. 9.10 **Global oil production, history so far plus two Hubbert's peak projections.**

The forecast for world petroleum depends entirely on what you call oil and how much is ultimately out there. Figure 9.10 shows world oil production in the past, fit to two bell curves, one for 200 Gton C in oil that will ultimately be extractable and one for 500 Gton C. We have already extracted 117 Gton C. If 500 Gton C is ultimately right, then the peak comes in the 2030s sometime. If 200 Gton C is right, consistent with BP, the peak should be now. The significance of the peak for the world economy is that demand and consumption of petroleum is growing exponentially, along with other social factors like population and GDP. If the constant or dwindling oil supply is unable to keep up with the exponentially rising demand, that's what they call shortage. The bottom line is that the time to watch out for is when the oil is half gone; things won't be fine until the last drop.

Natural gas is another basic type of fossil fuel. Natural gas comes out of the Earth as a gas instead of a liquid. It is mostly ***methane***, CH_4, with a few other slightly larger carbon molecules thrown in. Oil and oil source rocks ultimately convert to natural gas if they are buried deeply in the hot interior of the Earth. Methane can also be produced from organic carbon by bacteria. Methane sometimes comes out of oil wells, or it can be extracted from coal beds. Methane is more difficult to transport than liquid petroleum because it is a gas and must be held under pressure. For this reason, methane associated with oil in remote locations is sometimes just burned, giant dramatic flares into the sky, rather than collected. In Siberia, for example, natural gas is worth nothing. The ultimate extractable amount of natural gas is probably even more poorly known than it is for petroleum because it hasn't been explored for as intensively. Industry estimates are about 100 Gton C as natural gas in proven reserves. Just like for oil, the ultimately available natural gas reservoir will certainly be higher than that.

There is also a 363 kg gorilla in the picture, hiding in near-shore ocean sediments. ***Methane hydrates*** are methane gas molecules frozen into soccer-ball cages of water ice. Water can form these cages around any gas; all the gas is required to do is hold the little soccer-balls open. CO_2 forms hydrates on Mars. It turns out that there are thousands of Gtons of carbon in methane hydrate deposits on Earth, in mid-depth sediments of the ocean and in permafrost soils.

There are two required ingredients for methane hydrates: lots of methane and cool temperature. It is unusual to find methane hydrates very close to the sea floor because methane concentrations tend to be low in surface sediments. The temperature sets a lower limit to the depth range where methane hydrates are found because it gets hotter with depth in the Earth. Beneath the hydrate melting depth, methane often exists as bubbles of gas within the sediment. This layer of bubbles can be seen in sediments around the world in the reflection of seismic waves, echoes of explosions set behind ships. The reflective layer of bubbles tends to parallel the sediment surface, and so is called a ***bottom-simulating reflector***. The bottom-simulating reflector in seismic data is the best indication we have of how extensive the methane hydrate deposits are.

Methane clathrate deposits seem like the most unlikely and precarious of things. The ices themselves would float in seawater if they were not held down by the sediment above. The bubbles make the deposits even more unstable. They decrease the average density of the sediment column. If the mixture of solid, seawater, and bubbles (called a ***slurry***) starts to rise up, the gas bubbles would expand causing the slurry to rise even faster. Methane regularly escapes catastrophically from the sediment to the ocean, leaving behind explosion craters called ***pockmarks***. Methane may also be released by submarine landslides. We will hear a prognosis for methane clathrates in Chapter 12.

Returning to Fig. 9.1, we see how different countries get their energy. India and China, the largest of the developing world, use a lot of coal whereas Brazil uses mostly petroleum and hydroelectric power. Denmark is remarkable for windmills, which fall into the category "renewable" on this plot. France has invested heavily in nuclear energy (not renewable). The US energy sources are similar to the global average. In part this is because the United States is a big consumer of energy, accounting for about a quarter of energy use globally (Fig. 9.11).

If we divide the energy use of a country by the number of people, we get the energy use ***per capita***. We see in Fig. 9.11 that Americans use five times more energy than the global average citizen of the world, ten times more than the average Chinese or Indian, and twice as much even as the average European or Japanese.

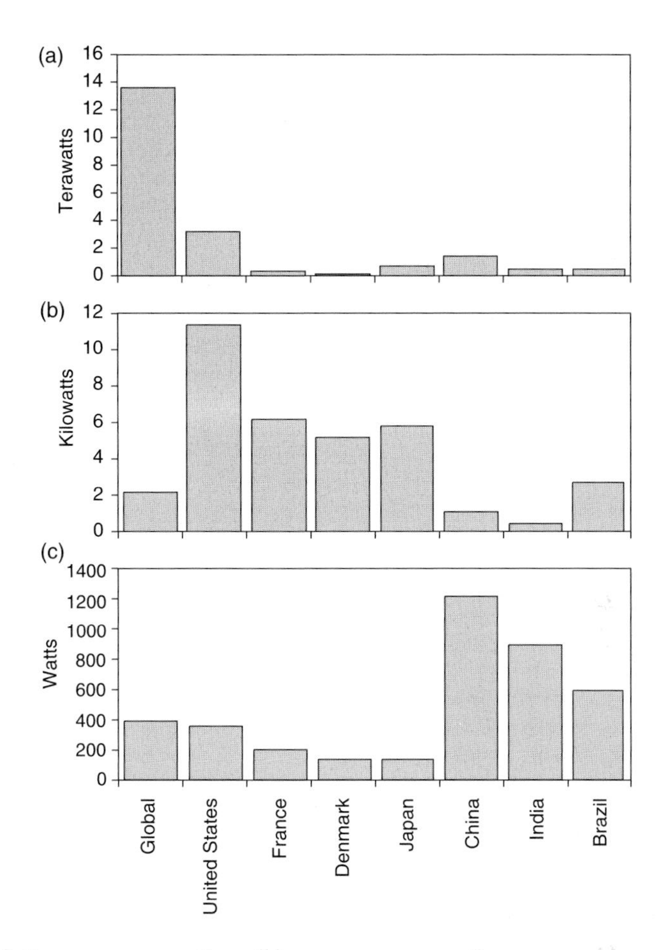

Fig. 9.11 **(a) Energy consumption, (b) energy consumption per person, and (c) energy consumption per dollar gross domestic product (GDP), globally and from various countries.**

Energy use is closely related to economic productivity, which is measured by the total value of everything produced for sale in a country in a year, the gross domestic productivity or ***GDP*** in units of dollars per year. We can divide energy use by GDP to derive the ***energy intensity***, a measure of the energy efficiency of an economy, not only reflecting waste but also the difference between heavy industry such as steel production versus high-tech industry. We see in Fig. 9.11 that the Europeans and the Japanese can make money using less energy than we can whereas the Chinese and Indians are more energy intensive (less efficient).

Carbon emissions from these countries look similar to the energy story (Fig. 9.12). The United States is responsible for about a quarter of global CO_2 emissions. Per person, the United States releases five times more than the global average, and more than twice as much as the Europeans or the Japanese, ten times more than the developing world. Per GDP, the United States is less than half as efficient as Europe or Japan.

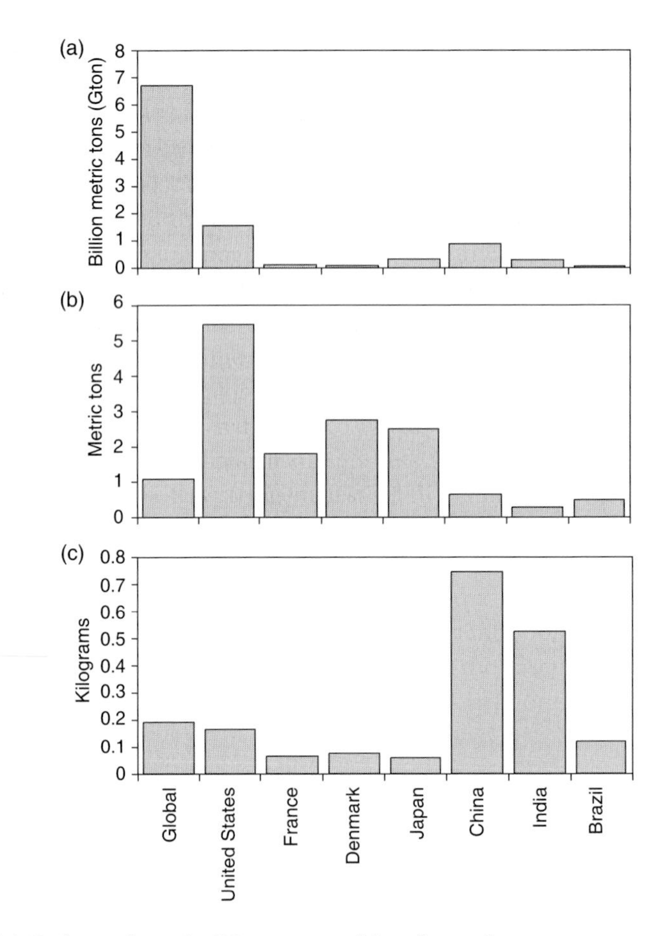

Fig. 9.12 **(a) Carbon release in CO_2 per year, (b) carbon release per person, and (c) carbon release per dollar gross domestic product, GDP, globally and from various countries.**

Future energy consumption

We have begun to see the factors that control CO_2 emission: population, GDP, energy use per dollar GDP, and carbon release per Watt of energy. CO_2 emission can be estimated by multiplying these factors together; check it out for yourself that the units cancel correctly. Constructing CO_2 emission from these pieces is called the Kaya identity; it looks like this.

$$CO_2 \text{ emission} = \text{population} \times \frac{\$\text{GDP}}{\text{person}} \times \frac{\text{Watts}}{\$} \times \frac{CO_2 \text{ emitted}}{\text{Watts}}$$

You can play with the Kaya identity and make your own forecast online at http://understandingtheforecast.org/models/kaya.html, and see results in Fig. 9.12.

Population has been rising at a rate of about 1.3% per year, but this rate of growth is not expected to continue. Rates of population growth tend to decrease with affluence of societies. Population is forecast to level off at some value, but that value could be

9 billion people or it could be 15 billion people. A typical "business-as-usual" forecast puts the leveling-off point at about 11 million people.

The second factor, \$GDP/person, has been rising throughout the last century at a rate of about 1.6% per year, rising from US \$930 in 1990 to \$4800 per person in 2000.

The third factor is the energy intensity: how many watts of energy it takes to make a dollar GDP. The energy intensity not only reflects efficiency but also the difference between heavy and light industry (Fig. 9.13c). Energy intensity has been dropping by a rate of about 0.55% per year over the past century, but the last few decades have seen a drop by about 1% per year. According to the model on the web site, the slower rate of decrease results in 825 ppm pCO_2 in 2100, while 1% per year results in about 700 ppm (Fig. 9.14).

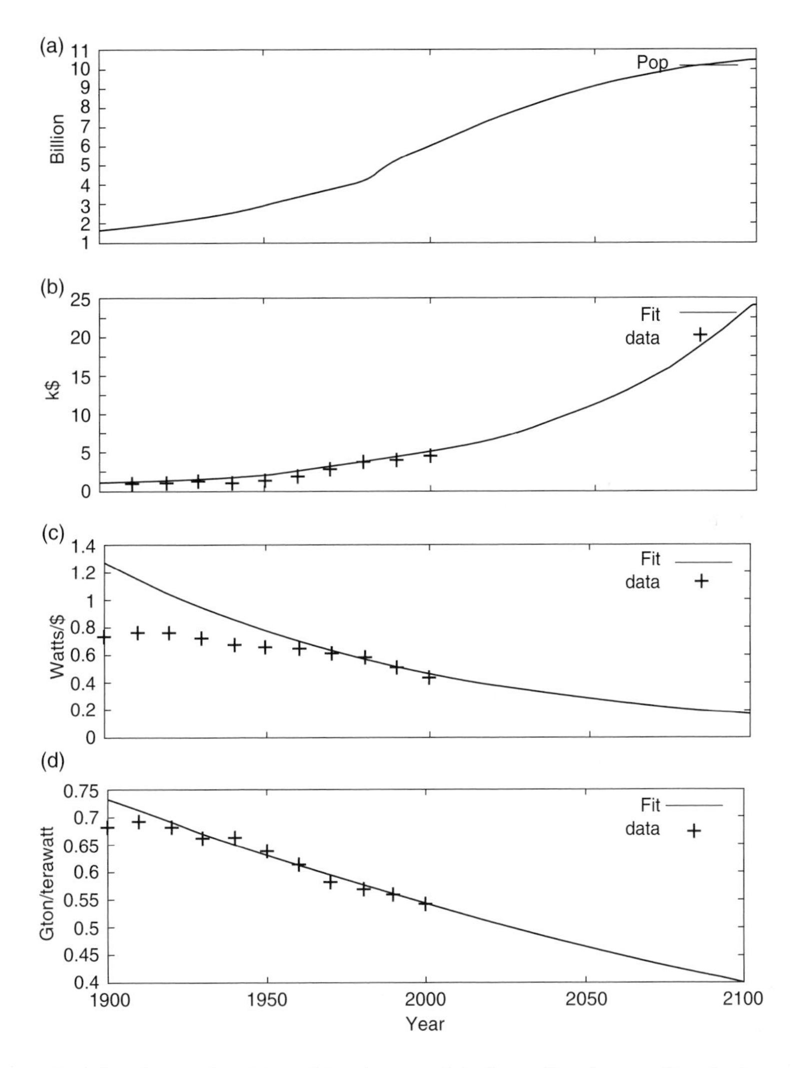

Fig. 9.13 **Results from the Kaya identity model (http://understandingtheforecast.org/Projects/kaya.html). (a) Population, (b) GDP per capita, (c) energy intensity, and (d) carbon efficiency.**

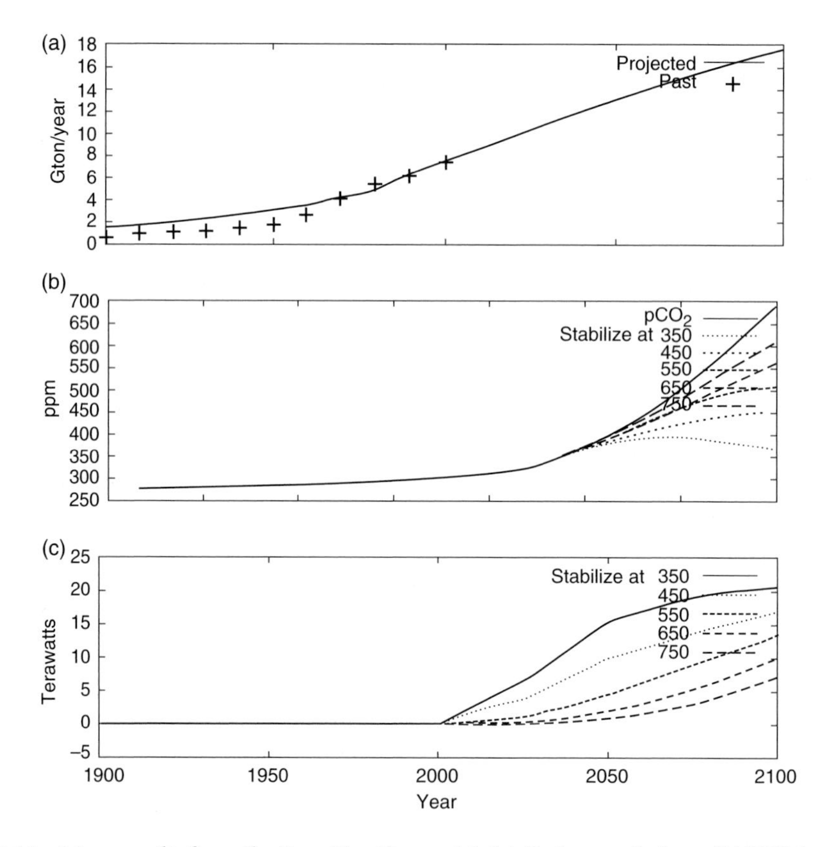

Fig. 9.14 **More results from the Kaya identity model. (a) Carbon emissions, (b) ISAM carbon cycle model pCO_2, and (c) carbon-free energy required for CO_2 stabilization.**

The final factor, the carbon released per energy yield, is not only the result of efficiency of the power plants but also the characteristics of the feedstock energy source. Coal contains intrinsically less energy per carbon than does oil or gas whereas nuclear and hydroelectric energy release no carbon at all. CO_2 emission scenarios from the ***Intergovernmental Panel on Climate Change***, or ***IPCC***, ranged from 6 to 26 Gton C emission per year in 2100. We will return to this model in Chapter 10 to see what happens to that CO_2 and in Chapter 13 to see what it means in terms of energy.

Take-home points

1. Ultimately, the energy available to mankind includes instantaneous solar energy, which is abundant but spread out; stored solar energy in the form of fossil fuels; and stored energy from stellar explosions in the form of radioactive uranium deposits.
2. Of the fossil fuels, coal is the most abundant. Oil may run out in the coming decades, and the peak rate of oil extraction may be upon us even now.

3. We can project energy demand in the future as the product of population, economic growth, and energy efficiency.

Projects

1. *Hubbert's peak.* Point your web browser to http://understandingtheforecast.org/Projects/hubbert.html.

    a. You will see two different data sets to plot against, along with the three parameters (knobs) that control the shape of the Hubbert curve. Nothing fancy here, we're just matching the curve to the data by eye. First start out with the US oil production. The page comes up with some values for the curve that look pretty good to me, but you should try varying the numbers in the box to see how tightly those values are constrained. In particular, there may be combinations of values that could be changed together to fit the data nearly as well, but with different values. How much wiggle room is there for US oil production?
    b. Now switch to global oil production with the pull-down menu. When do you forecast the peak of oil extraction? How does it depend on your assumption of how much oil will eventually be extractable? How much wiggle-room is there for the year of the peak in global oil extraction?

2. *The Kaya identity.* Point your web browser to http://understandingtheforecast.org/Projects/kaya.html.

    a. Find the plots for GDP per capita, energy intensity, and carbon efficiency, and compare the model hindcast (the solid line) with the data (plusses). How well constrained are the growth rates by the past data? Of course, the future may not follow the dictates of the past; this is not a mechanistic prediction but just a blind extrapolation. Using the past as a fallible guide, however, take a guess at what the range of possibilities is for each of the input values.
    b. How much carbon is mankind predicted to emit by the end of the century? Using the uncertainty ranges you made in 2a, what is the highest and lowest plausible carbon emission for 2100?

3. *IPCC CO_2 emission scenarios.* Open a new browser window for the Integrated Science Assessment Model (ISAM) carbon cycle model, at http://understandingtheforecast.org/Projects/isam.html. This page shows the results of IPCC carbon emission scenarios made by more sophisticated crystal balls than our simple Kaya identity (and then offers to run the results of the scenarios through a carbon cycle model). On the Kaya page, try to reproduce the 2100-year carbon emissions from scenarios A (business-as-usual, BAU), B (BAU with carbon stabilization), C (slow growth), and F (gonzo emissions). What input parameters are required?

Further reading

British Petroleum, Energy in focus, *BP Statistical Review of World Energy*, 2004.
Deffeyes, Kenneth S., *Hubbert's Peak: The Impending World Oil Shortage*, Princeton, 2003.
International Energy Outlook, 2003. Energy Information Administration, US Department of Energy. www.eia.doe.gov.
Rogner, H.-H., An assessment of world hydrocarbon resources, *Annual Review of Energy and the Environment* (1997), *22*, 217–62.

10
The perturbed carbon cycle

We describe three geochemical global change issues here. Ozone in the stratosphere is beneficial because it shields us from ultraviolet (UV) sunlight, but it is depleted by chlorofluorocarbons. Ozone in the troposphere is harmful to lungs and to plants, but it is produced by automobile emissions. Neither problem has much to do with global warming but we present it to allay confusion about this.

Methane is produced by livestock and rice paddies as well as natural sources. The lifetime of methane in the atmosphere is about a decade. The concentration of methane in the atmosphere can be calculated from the source flux multiplied by the lifetime.

CO_2 is accumulating more slowly in the atmosphere than we are releasing it because of natural uptake by the terrestrial biosphere and the ocean. Ocean uptake is easier to quantify than terrestrial because chemical properties in the ocean are less patchy than they are on land. CO_2 has a long lifetime in the atmosphere, and will influence the climate for hundreds of thousands of years into the future. Long-term projections of carbon use, compared with the demands of stabilizing the CO_2 concentration of the atmosphere, indicate that a major new source of carbon-free energy will be required in the coming century, with the capacity to produce as much energy as we are currently consuming or even more.

Ozone

Let's get one issue out in the open and out of the way before we begin thinking about global warming. The ozone hole is not the same as global warming. ***Ozone*** is a reactive oxygen molecule comprised of three oxygen atoms. Ozone in the stratosphere is produced as O_2 molecules which are zapped by energetic UV-C light, breaking apart into two very reactive oxygen atoms. Each of these may find another O_2 molecule and join it, to form O_3, ozone. Ozone itself absorbs UV light called UV-B that is less energetic but more abundant than the UV-C required to break up an O_2 molecule. Stratospheric ozone filters UV-B radiation that might otherwise reach the surface, causing skin cancers and sunburn.

Atmospheric chemists predicted that ozone concentrations in the stratosphere might be depleted by chemical reactions accompanying the breakdown of chemicals called chlorofluorocarbons or freons, inert chlorine-bearing compounds that are used in refrigerators and air conditioners. Their prediction was that stratospheric ozone concentrations would gradually decrease over the decades. This has been observed, but what scientists did not anticipate was a surprise called the ***ozone hole***. The ozone hole is located in the southern hemisphere over Antarctica. In this region, during winter time it gets cold enough that an unusual form of cloud forms in the stratosphere, comprised of frozen particles of nitric acid, HNO_3. It turns out that the frozen nitric acid clouds convert the chlorine from the breakdown of freons into a very reactive form that doesn't just deplete ozone by a small amount, but consumes it entirely within that air mass. The ozone hole was first observed in measurements of ozone concentration made by hand, on the ground. After these measurements, old satellite ozone records were examined, and it turned out that the satellite had been seeing the ozone hole for several years, but a data quality algorithm had been programmed to throw out any data which violated common sense, and out with the bad data went the ozone hole. This is a measure of how much of a surprise it was.

The ozone hole is a terrific example of a ***smoking gun***, clear-cut proof of the detrimental effects of chlorofluorocarbons on stratospheric ozone. Before the ozone hole was detected, the models predicted a gradual decrease in stratospheric ozone globally, superimposed on natural variability; a difficult prediction to test. As a result of the ozone hole, and the availability of economically and technologically viable alternatives, many countries of the world ratified the Montreal Protocol in 1987, phasing out further production and release of chlorofluorocarbons to the environment.

Ozone is also produced by reaction of gases from industrial activity, mostly automobile exhaust, in surface urban smog. The ingredients for ozone production in urban air are evaporated organic carbon compounds, like evaporated gasoline or molecules that are emitted naturally from trees and plants, nitrogen oxide compounds that are emitted from automobiles, and sunlight. When ozone concentrations exceed a toxic limit, asthma sufferers begin to feel discomfort, and plant leaves get "burned" and scarred.

Ozone in the stratosphere interacts with climate change caused by greenhouse gases such as CO_2. The main role that ozone plays in the energy budget of stratospheric air is it absorbs UV light from the Sun, heating up the air. Ozone is the reason why there is a stratosphere. If ozone did not heat up the stratosphere, the atmosphere would continue to get colder with altitude as it does in the troposphere. The stratosphere would disappear into the troposphere. Decreasing ozone results in less heating and cooler air in the stratosphere.

Rising CO_2 concentrations also cool the stratosphere. The role that CO_2 plays in the heat balance of stratospheric air is that it acts like a radiator fin, exporting heat energy as IR light. For this reason, an increase in CO_2 concentration in stratospheric air causes the temperature there to go down. Ozone is also a greenhouse gas, but it has a stronger influence on the heat balance by absorbing UV.

So we have stratospheric cooling for two reasons: falling ozone concentration and rising CO_2. Now the thread of the story makes its way back to ozone. The stratospheric clouds that cause the ozone wipe-out in the ozone hole are more prevalent when the air is cold. They are clouds of nitric acid ice, which only form in extremely cold air. Decreasing ozone leads to decreasing stratospheric temperature, which leads to further ozone depletion. This is an example of a positive feedback.

Ozone is confusing to a student of the environment because in the stratosphere it is a good thing but industrial activity is acting to deplete it, whereas in the troposphere it is a bad thing and industrial activity tends to produce it.

Table 10.1 Natural and anthropogenic sources of methane

Source	Gton C/year
Natural sources	
Wetlands	0.075–0.15
Termites	0.015
Human sources	
Energy	0.075
Landfills	0.03
Ruminant animals	0.075
Rice agriculture	0.05
Biomass burning	0.03

Methane

Methane is a greenhouse gas, 20 times more powerful per molecule than CO_2 at current concentrations (see Chapter 4). Methane has natural sources as well as additional anthropogenic sources to the atmosphere (Table 10.1). Once released into the atmosphere, methane reacts slowly with activated oxygen compounds to oxidize back to CO_2. The reactive oxygen compounds are produced by sunlight. In the absence of sunlight methane and O_2 gas coexist in ice core bubbles for hundreds of thousands of years with no reaction. Put it in the sunlight, and it slowly burns up.

The rate of methane emission and the atmospheric lifetime of methane together determine the concentration of methane in the atmosphere. Let's assume that the methane release rate is steady from one year to the next for a long time. Let's also assume that the emission of methane must be balanced by the rate of methane decomposition. The fluxes balance exactly if the methane concentration were constant with time. As it is, methane is rising with time, so our steady-state assumption is not strictly correct, but it is close enough to be useful. The steady-state assumption is

$$\text{Emission [Gton C/year]} = \text{Decomposition [Gton C/year]}$$

The lifetime of a methane molecule in the present-day atmosphere is about a decade. Methane is consumed by reactive oxygen compounds in the atmosphere, in particular, a molecule called OH radical. OH radical is related to ozone, so one could imagine a change in OH radical driven by the change in ozone chemistry of the atmosphere. It could be that if the methane concentration were higher, the lifetime might be longer because degradation might be limited by the availability of OH radical molecules. If however we toss out these potential complications, and just assume that the atmospheric lifetime of methane is and will always be a decade, we could just write that the degradation rate of methane is

$$\text{Decomposition [Gton C/year]} = \text{Inventory [Gton C]/Lifetime [years]}$$

Combining these two equations, we get

$$\text{Emission [Gton C/year]} = \text{Inventory [Gton C]/Lifetime [years]}$$

Rearranging,

$$\text{Inventory [Gton C]} = \text{CH}_4 \text{ emission [Gton C/year]} * \text{Lifetime [years]}$$

This relation tells us that the methane concentration in the atmosphere is linearly related to the methane source to the atmosphere, as long as the lifetime of methane stays the same. If we doubled the emission, after a few decades, the steady-state concentration would double. The real world may be a bit more complicated because the lifetime might change as the methane concentration goes up.

One of the natural sources of methane to the atmosphere is the degradation of organic carbon in freshwater swamps. Organic carbon degrades first by reaction with O_2, as we have discussed in Chapter 8. In seawater, after the O_2 is gone, organic carbon reacts with sulfate ion, SO_4^{2-}, to produce hydrogen sulfide, H_2S. After oxygen and sulfate are depleted, methane is produced from organic carbon by fermentation. This is how the methane is produced that freezes into clathrate deposits below the sea floor.

In freshwater, there is not much SO_4^{2-}, so as soon as oxygen is gone, methane production begins. Methane is found much shallower in freshwater mud than in salt water mud. If you step in mucky swampy freshwater mud, you may see bubbles of methane rising up around your legs. Methane is sometimes referred to as swamp gas for this reason, and it is one of the usual suspects blamed for sightings of UFOs, flying saucers. Bubbles of ancient atmosphere preserved in ice cores tell us that the methane concentration has fluctuated with climate state over the past 400,000 years (the longest ice core yet available), with lower methane concentrations during colder, drier climate stages (Fig. 8.3). This is interpreted to be the result of the changing abundance of swamps.

Anthropogenic sources of methane include production in the guts of ruminant animals, and release as leakage by the fossil fuel industry. Rice paddies often provide ideal anoxic freshwater environments for methane production. The methane concentration has doubled over its preanthropogenic concentration (Fig. 10.1), and is responsible for a quarter of anthropogenic greenhouse heat trapping (Fig. 10.2).

CO_2

There are two main anthropogenic sources of CO_2 to the atmosphere. One is deforestation. A heavily wooded forest holds more carbon per area than does a plowed agricultural field. The world's forests started to feel the axe thousands of years ago with the development of agriculture and the growing of the human population. Most of the temperate latitudes have been cut long since, and the tropics are currently being cut. The year 1750 has been taken as the beginning of the anthropogenic CO_2 rise, although it has been argued that both CO_2 and methane may have started rising from their "natural" trajectories thousands of years ago. The rise in atmospheric CO_2 after

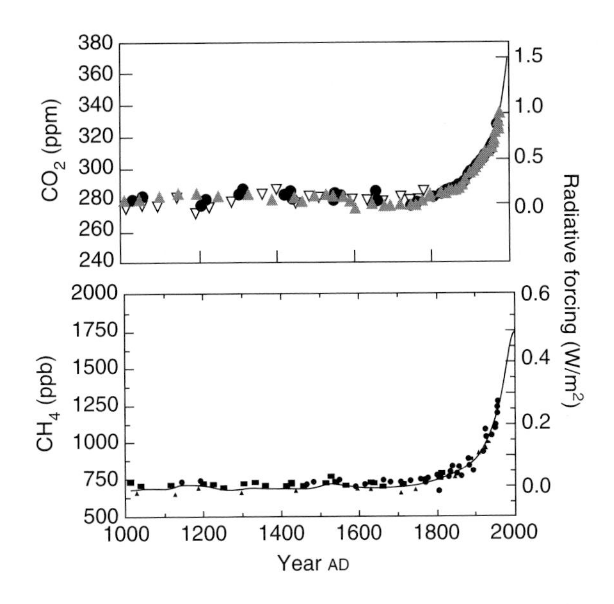

Fig. 10.1 **History of CO_2 and CH_4 concentrations in the atmosphere, from ice cores (symbols) and atmospheric measurements (solid lines). Replotted from IPCC (2001).**

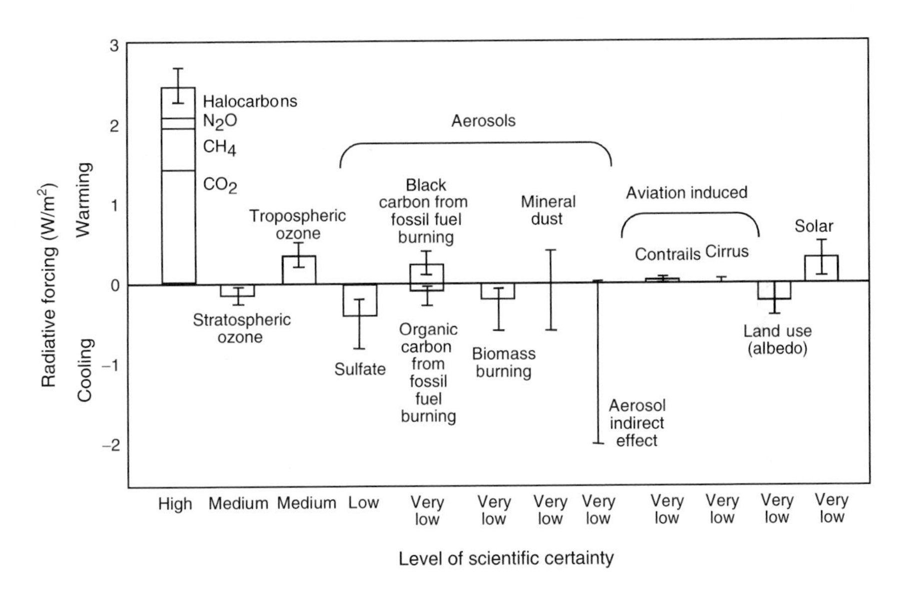

Fig. 10.2 **The impact of various human-related climate drivers on the energy budget of the Earth, in watts per square meter, relative to the year 1750. Replotted from IPCC (2001).**

1750 (Fig. 10.1) is clearly the result of deforestation, perhaps in the New World (the "pioneer effect"). Today, visible deforestation is mostly to be found in the tropics, and accounts for about 2 Gton C/year release of CO_2 to the atmosphere.

The other main anthropogenic CO_2 source is of course the combustion of fossil fuels discussed in the last chapter. Fossil fuel combustion releases 5 Gton C/year, rising exponentially, driven by population growth and economic growth, rising in spite of increases in energy and carbon fuel efficiency.

Combining fossil fuel combustion and tropical deforestation, mankind is releasing carbon to the atmosphere at a rate of about 7 Gton C/year. The atmospheric CO_2 inventory is rising at a rate of about 3 Gton C/year. Where is the other 4 Gton C/year? There are two main natural sinks for CO_2 that are operating today; one is the oceans and the other is the terrestrial biosphere.

Carbon uptake by the terrestrial biosphere on land, the ***terrestrial carbon sink***, is difficult to measure. In comparison with the ocean, the distribution of carbon on land is very spotty. Recall from Chapter 7 that most of the carbon on land is in the soil, rather than in the trees where we could see it. In soils, the amount of carbon depends on the recent history of the land: fires, agriculture, erosion, and so on. It is difficult to know precisely how much carbon there is on land because the measurements are so variable; you would have to make a lot of measurements in order to average out all the noise of natural variations. As a result of this, it would be possible to increase the amount of carbon on land, a little bit here or there, in a way that would be entirely invisible to direct measurements. The land is playing two roles in the carbon budget story, one as a visible deforestation source and another as a potential invisible carbon uptake sink.

One way to estimate the invisible terrestrial uptake is by putting together the rest of the carbon budget and assign the terrestrial biosphere whatever is left over. The chemical properties of seawater vary more smoothly than they do on land, so our estimates of ocean uptake are better than they are for the land. Another is to measure CO_2 concentrations in the atmosphere, in the winds as they blow across the land, to see if CO_2 is going into or coming out of a given forest. This doesn't sound easy, but it can be done. There is a network of "CO_2 observatories" around the globe, where precise CO_2 concentration measurements are made daily, for uses such as this.

There are several reasons why the land may be willing to take up CO_2 as the atmospheric CO_2 concentration rises. One is that with warming, there will be a longer growing season. This has been observed in many climate and botanical records. With warming, some tundra areas become amenable to conquest by forests. Rising CO_2 in the atmosphere may also directly encourage plants to grow faster by a process known as ***CO_2 fertilization***. Plants run their photosynthetic machinery inside waxy walls on the surfaces of leaves. Gases are exchanged with the outside atmosphere through adjustable vents called ***stomata***. When the leaf needs CO_2 for photosynthesis, the stomata open. The cost of opening stomata, though, is loss of water. So if CO_2 concentrations were higher in the outside atmosphere, plants could get the CO_2 they need without opening their stomata as much or as often. They could therefore be stingier with their water. There is no doubt that this is a real effect; CO_2 concentrations in greenhouses are typically higher than in the outside atmosphere, one of the ways that greenhouses are good for plants. However, in the real world, plant growth is very often limited by something else other than water stress, such as fertilizers like nitrogen or phosphorus. Scientists do CO_2 fertilization experiments in natural settings by pumping CO_2 continuously into the air. When the wind changes, they adjust the location of the CO_2 vent, so that the target grove is always downwind from a CO_2 source. These experiments go on for years!

What they tend to find is an initial growth spurt from CO_2 fertilization, followed by a leveling off at something like the initial rates.

There is another process that might affect CO_2 storage on land, which is temperature sensitivity to respiration, the process that converts soil organic carbon back into CO_2. Soil respiration really gets going as it gets warmer. Think of a ham sandwich, half of which is safely stowed in the refrigerator while the other half sits on a plate in the Sun. Which half will stay tasty longer? For this reason, there is very little organic matter in tropical soils, while high latitudes may host peat deposits that contain prodigious carbon deposits for thousands of years. Warming and melting and decomposition of high-latitude permafrost may contribute CO_2 to the atmosphere.

Uptake of fossil fuel CO_2 by the oceans is called the ***ocean carbon sink***. The ocean sink depends on ocean circulation, and on the chemical forms that dissolved CO_2 takes in seawater. The ocean covers 70% of the Earth's surface, and the length and width of the ocean are huge compared with its depth, which averages about 4 km. The deep ocean is so close to the surface, and yet it is so very far away. The way the ocean circulates, the deep ocean is very cold, and the only place where surface waters are cold enough to mix with the deep ocean is in high latitudes. The ocean surface is huge, but the deep ocean, which is the largest water type in the ocean, only sees the atmosphere through a very small area of sea surface.

The densest water at the sea surface is in the Antarctic and in the North Atlantic because it is cold there. Surface waters from these locations sink to the deep ocean, filling up the entire deep ocean like a bucket with cold polar water that is only a few degrees warmer than freezing. The cold deep ocean fills up until cold polar waters underlie the warm surface waters in lower latitudes. The warmer waters mix with the cooler, eroding the cold water and making room for more of the coldest water to continue filling the deep sea. As new cold water flows from the high-latitude surface ocean into the abyss, it carries with it atmospheric gases like anthropogenic CO_2, a process known as ***ocean ventilation***. It takes centuries for the waters of the deep ocean to travel through this cycle. For this reason, the timescale for getting anthropogenic CO_2 into the deep ocean is centuries.

The shallower ocean has other water masses and circulation modes that are wondrous to learn about and study if one is of a mind to. The zone of the ocean separating the warm from the cold is called the ***thermocline***. Thermocline waters may be exposed to the atmosphere in winter, when the sea surface waters are cold. Once a parcel of thermocline water becomes isolated from the sea surface, it follows a trajectory determined by its density and by the rotation of the Earth. Thermocline waters ventilate to the atmosphere on a timescale of decades.

The surface ocean water mass is not as large as the deep sea or the thermocline, but it is a respectable carbon reservoir of its own. Turbulence generated by the wind acts to mix the surface ocean down to a typical depth of 100 m. For most gases, this surface ocean layer 100 m thick would equilibrate with the atmosphere in about a month.

CO_2 differs from other gases, however, in that it has chemical equilibrium reactions with water and hydrogen ions. Hydrogen ions are very reactive. If a solution has a high concentration of hydrogen ions, we call it ***acidic***. A strongly acidic solution, such as battery acid for example, can burn your skin or clothes by chemical reaction with hydrogen ions. The acidity of a solution is described by a number called the *pH* of the solution, which can be calculated as

$$pH = -\log_{10} [H^+]$$

The $\log_{10}$ is the base-10 logarithm, meaning that if $x = 10^y$, then $\log_{10} y = x$. The hydrogen ion concentration is denoted by the square brackets, and is expressed in units of moles of H^+ per liter of solution. A ***mole*** is simply a set number of atoms or molecules called the Avogadro's number and is equal to $6.023 \cdot 10^{23}$. The hydrogen ion concentration in seawater usually ranges from 10^{-7} to $10^{-8.3}$ mol of H^+ per liter. The pH of seawater therefore ranges from 7 to 8.3. Note that the more acidic the solution, the lower the pH of the solution. Ads for shampoo claim "low pH" as though pH were some toxic ingredient. I guess it sounded better than calling the shampoo "strongly acidic."

When CO_2 dissolves in water, it reacts with water to form ***carbonic acid***, H_2CO_3.

$$CO_2 + H_2O \Leftrightarrow H_2CO_3 \tag{10.1}$$

Carbonic acid loses a hydrogen ion (that's what acids do, in general; release hydrogen ions) to form ***bicarbonate*** ion, HCO_3^-

$$H_2CO_3 \Leftrightarrow H^+ + HCO_3^- \tag{10.2}$$

A second hydrogen ion can be released to form ***carbonate ion***, $CO_3^=$

$$HCO_3^- \Leftrightarrow H^+ + CO_3^= \tag{10.3}$$

The concentrations of carbonic acid, bicarbonate, and carbonate ions control the acidity of the ocean, just as they control the acidity of our blood and cell plasma.

These chemical reactions are fast enough that the distribution of chemical species will always be in their lowest energy distribution. Many chemical reactions that we will encounter will not be in equilibrium, so we should enjoy this equilibrium system while we have it. Equilibrium reactions can be easily and very precisely predicted. In a qualitative way, we can use an idea known as ***le Chatelier's principle*** to take a stab at the behavior of an equilibrium system. Le Chatelier's principle states that an addition or removal of a chemical on one side of the chemical equilibrium will cause the reaction to run in the opposite direction to compensate for the change. Take some of something out, the equilibrium will put some of the something back. Add more something, the equilibrium will remove some of the something.

Le Chatelier's principle is treacherous for students of the carbon system in seawater, though, because we have an innate human tendency, I have found, to ignore the hydrogen ions. They are such tiny things, after all. There are far fewer hydrogen ions in seawater than there are of the dissolved carbon species. What this means, however, is that a small change in the concentrations of the carbonate species might make a huge change in the hydrogen ion concentration. The safest assumption to make is that the carbon species have to get along together without counting on sloughing off too many hydrogen ions at all. Hydrogen ion is such a tiny slush fund, it might as well not exist. We can combine reactions (10.1–10.3) into a single reaction, in such a way that we don't allow any production or consumption of hydrogen ions:

$$CO_2 + CO_3^{=} + H_2O \Leftrightarrow 2HCO_3^{-} \qquad (10.4)$$

To this reaction we can apply le Chatelier principle with impunity. If we were to add CO_2 to this system, the equilibrium would compensate somewhat by shifting to the right, consuming some of the CO_2 by reacting it with $CO_3^{=}$.

Seawater has the capacity to absorb or release more CO_2 than it would if CO_2 had no pH chemistry because of the other carbon reservoirs HCO_3^{-} and $CO_3^{=}$. It is like sitting at a poker game with a rich uncle sitting behind you covering most of your losses and taking splits of your winnings. The pH reactions of H_2CO_3, HCO_3^{-}, and $CO_3^{=}$ are called a ***buffer*** because any changes to the chemistry tend to be resisted or buffered by the chemical reactions. The CO_2 concentration of seawater is buffered by its pH chemistry. Another way of saying this is that the seawater has greater capacity to hold carbon than it would if CO_2 were not buffered.

The strength of the buffer is about a factor of 10, meaning that seawater has a capacity to hold 10 times as much CO_2 as it would if there were no buffer chemistry. The factor of 10 comes from the fact that there is about 10 times more $CO_3^{=}$ than dissolved CO_2 in surface ocean water. Carbonate ion is our anti-CO_2, reacting with new CO_2, hiding it away as bicarbonate; this is the action of the buffer.

As the CO_2 concentration in the atmosphere increases, and CO_2 invades the ocean, the concentration of carbonate ion goes down, according to the equilibrium chemical reaction (10.4). As the carbonate ion becomes depleted, so is its ability to buffer CO_2. Ocean uptake of new CO_2 would decrease as a result of this. The future of ocean uptake of fossil fuel CO_2 may also be affected by changes in the circulation of the ocean. Surface warming is expected to be most intense in high latitudes because of the ice albedo feedback (Chapter 7). If the high latitudes warm, the overall circulation of the subsurface ocean may decrease. The circulation of the ocean may stagnate, slowing uptake of CO_2.

Biology in the ocean acts to decrease the CO_2 concentration of surface waters by converting CO_2 into organic carbon via photosynthesis (Chapter 8). Dead phytoplankton sink from surface waters exporting their carbon to the deep sea. This processes has been termed the ***biological pump***. If all the life in the ocean were killed, that is, if the biological pump were stopped, then the CO_2 concentration of the atmosphere would rise. If the biological pump were stimulated to work harder, it could decrease the CO_2 concentration of the atmosphere. One proposal for dealing with global warming is to fertilize the ocean with iron. Iron concentrations are extremely low in remote parts of the ocean, far away from iron deposition from dust and iron bleeding from surface sediments. Supplying iron to the phytoplankton has been shown to stimulate phytoplankton growth in the Southern Ocean around Antarctica for example. The problem is that it takes hundreds of years for the ocean and the atmosphere to negotiate what the atmospheric CO_2 concentration should be; it's slow, recall, because the ocean circulation is so slow. Model studies have shown that fertilizing the Southern Ocean for hundreds of years might bring the CO_2 concentration of the atmosphere down, but fertilizing for a few decades has very little impact.

Decreasing carbonate ion may also be detrimental to coral reef and other organisms that produce limestone, $CaCO_3$, from calcium ion, Ca^{2+}, and carbonate ion. It's like pouring vinegar on limestone steps; you will see bubbles as the acid of the vinegar converts the $CaCO_3$ to CO_2. Fossil fuel CO_2 is itself an acid, and drives $CaCO_3$ to dissolve. Note the counterintuitive reverse behavior; one might have thought that adding CO_2 to the oceans would lead to an increase in the amount of carbon that winds up as $CaCO_3$. The response is opposite this expectation because CO_2 is an acid and $CaCO_3$ reacts with acid. Fish and other aquatic organisms also react poorly to the acidity and higher CO_2 concentrations resulting from fossil fuel CO_2 release. This danger is called the ***acid ocean***.

Uptake of CO_2 into the oceans has been estimated by a number of different independent methods. These include measurements of the chemical concentrations throughout the ocean, and modeling the circulation and carbon cycle. Other chemicals serve as tracers for how the ocean circulates. These include radioactive elements produced naturally by cosmic rays, or in nuclear bomb tests in the 1960s, and industrial chemicals like chlorofluorocarbons. Another distinction between dissolution of CO_2 in the ocean versus uptake of CO_2 by photosynthesis on land is that photosynthesis releases oxygen, whereas dissolution in water does not. So you can measure the change in CO_2 and O_2 in the atmosphere to figure out what fraction of the missing CO_2 is going into the ocean versus into the terrestrial biosphere by photosynthesis.

There are discrepancies between the different estimates of ocean and land carbon uptake, and uncertainties associated with each method, but in general they all point to about a 50 : 50 split. The ocean gets about 2 Gton C/year of the anthropogenic CO_2 and the terrestrial biosphere gets 2 Gton C/year. The terrestrial uptake of new CO_2 is thought to be taking place in high northern latitudes, perhaps into the great forests of Canada and Siberia.

What about carbon uptake by the natural world in the future? It is certainly possible that the land carbon reservoir will change in the coming centuries, depending on how people decide to use the land surface, in addition to biological factors such as CO_2 fertilization and increases in soil respiration. So far the amount of carbon released, about 300 Gton C, is smaller than the size of the terrestrial biosphere (500 Gton C), especially if we consider soil carbon (1500 Gton C). It is reasonable to think about the land surface as a potential major player in the carbon budget. However, if we combust all of the coal or methane clathrate deposits, the amount of CO_2 we could release could be 5000 Gton C or more, several times larger than the carbon stored on land. It would be difficult to imagine the terrestrial biosphere saving the day in that case.

After hundreds of years, about 75% of the fossil fuel CO_2 will dissolve in the oceans, while the remaining 25% remains in the atmosphere, awaiting slow chemical reactions with rocks that will ultimately consume it (Fig. 10.3). The CO_2 invasion lowers the pH of the ocean and the concentration of carbonate ion. On timescales of thousands of years, the pH and carbonate ion concentration of the ocean are controlled by limestone,

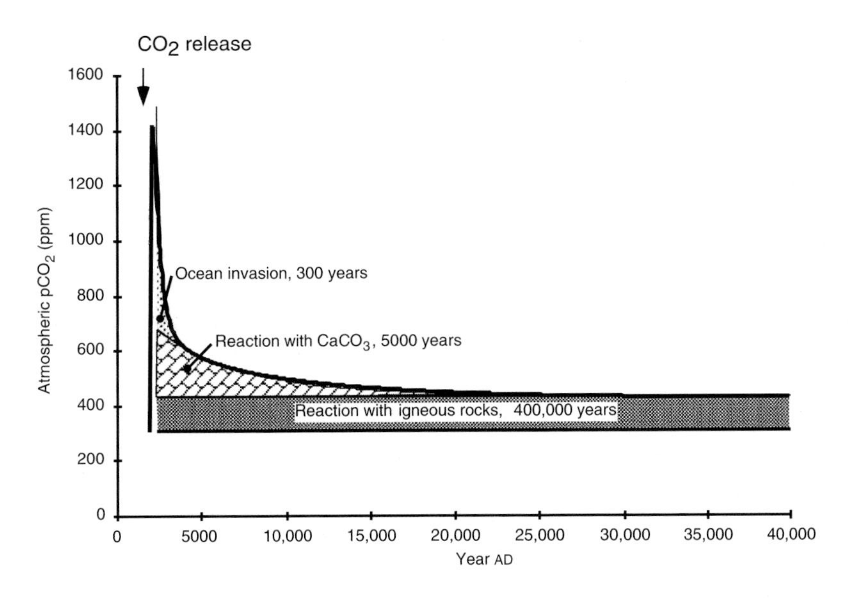

Fig. 10.3 **Long-term fate of fossil fuel CO_2. Reprinted from Archer JGR (2005).**

$CaCO_3$. Limestone on land dissolves, a chemical reaction we have already defined as weathering. Dissolved limestone flows to the ocean in rivers. Plankton re-form the solid limestone to make shells of $CaCO_3$, some of which sink to the sea floor and are buried.

The effect of the CO_2 invasion of the ocean will be to make it harder for little shells to be buried. In fact, if we ultimately release 1000 Gton C or so, there will be net dissolution of $CaCO_3$ from the sea floor. Mankind will have reversed the net sedimentation of the ocean! Engineers in Chicago early in the last century reversed the flow direction of the Chicago River. That was impressive in its time, but it was nothing compared to this. The rate of $CaCO_3$ weathering will exceed $CaCO_3$ burial, and so dissolved $CaCO_3$ will accumulate in the ocean. The fossil fuel CO_2 acts as an acid, lowering the pH of the ocean, while the dissolved $CaCO_3$ is a base, pushing ocean pH back up toward its natural value. Restoring the pH of the ocean will draw down the atmospheric CO_2 somewhat. This process will take thousands of years.

On timescales of hundreds of thousands of years, the silicate weathering thermostat, defined and described in Chapter 7, will act to pull CO_2 down the rest of the way toward the pre-anthropogenic value. The bottom line is that about 15–30% of the CO_2 released by burning fossil fuel will still be in the atmosphere in 1000 years, and 7% will remain after 100,000 years. Truly, global warming is forever.

Of the 7 Gton C/year that mankind is releasing to the atmosphere today, 4 Gton C/year is going away as quickly as we release it. This leads to a simple but powerful conclusion: if we want atmospheric CO_2 to stop going up, tomorrow, we have to reduce our CO_2 emissions from 7 Gton C/year down to 4 Gton C/year, say a reduction of total carbon emission by 40%. Then the CO_2 concentration in the atmosphere would stop rising, but remain at its current level of 365 ppm. This could continue until the terrestrial biosphere and the ocean equilibrated at this new level, "filled up" with the new higher CO_2. No one has any idea how long it would take for the terrestrial biosphere to saturate or fill up, but the ocean would take several centuries. If emissions were stopped after that, the atmospheric CO_2 concentration would remain at 365 ppm for thousands of years.

The aim of the Kyoto Protocol, the international agreement to reduce CO_2 emissions discussed in Chapter 13, is to reduce emissions to about 6% below the 1990 levels, resulting in emissions that are still very close to 7 Gton C/year. CO_2 emissions under business-as-usual are projected to grow, so the rather modest-sounding 6%-below-1990 target actually amounts to about 30% reductions from the projected 2010 rate. Still, this is just a drop in the bucket of what would be required to truly stabilize the CO_2 concentration of the atmosphere. The Kyoto protocol by itself is not sufficient to end the problem of global warming; it can only be the first step, alas.

Take-home points

1. The ozone hole is not global warming. They are different issues.
2. Methane has a short lifetime in the atmosphere.
3. CO_2 has a long lifetime in the atmosphere. Stabilizing CO_2 in the atmosphere at some "safe level" (whatever that is) will require major new energy initiatives.

Projects

1. *Long-term fate of fossil fuel CO_2.* Use the online geologic carbon cycle model at http://understandingtheforecast.org/Projects/geocarb.html. Use the default setup of the model, and notice that the CO_2 weathering rates etc. for the transient state are the same as for the spinup state. So if there were no CO_2 spike at all, there would be no change in anything at year 0. (Go ahead, make sure I'm not lying about this.) Release some CO_2 in a transition spike, 1000 Gton or more or less, and see how long it takes for the CO_2 to decrease to a plateau. There are two CO_2 plots in the output, one covering 100,000 years and one covering 2.5 million years. How long does it take for CO_2 to level out after the spike, according to both plots?
2. *Effect of cutting carbon emissions.* Look at the online ISAM global warming model at http://understandingtheforecast.org/Projects/isam.html.

   a. Run the model for the "Business-as-usual" case (Scenario A), note the pCO_2 concentration of the atmosphere in the year 2100.
   b. Estimate the decrease in fossil fuel CO_2 emission that would be required to halt the increase in atmospheric CO_2, based on the present-day CO_2 fluxes into the ocean and into the terrestrial biosphere. Test your prediction by imposing these fluxes from the present-day onward to the year 2100.

c. Repeat experiment 2b but delaying the cuts in fossil fuel emissions to the year 2050. What is the impact this time, on the maximum pCO_2 value we will see in the coming century?

Further reading

Online Trends, a Compendium of Data of Global Change, from the Carbon Dioxide Information Analysis Center, Oak Ridge National Laboratory. http://cdiac.esd.ornl.gov/trends/trends.htm.

The Discovery of Global Warming (Harvard University Press, 2003) by Spencer Weart. The history book again. Read about early measurements of atmospheric CO_2.

IPCC Scientific Assessment 2001, from Cambridge University Press or downloadable from http://www.grida.no/climate/ipcc_tar/. Chapter 3, *The Carbon Cycle and Atmospheric Carbon Dioxide.*

Archer, D., Fate of fossil-fuel CO_2 in geologic time, *J. Geophys. Res. Oceans*, doi:10.1029/2004JC002625, 2005.

Part III

The forecast

11
Is it reliable?

The temperature history of the last century is known from a variety of independent sources that agree with each other pretty well. Reconstructions of the temperature over the last thousand years show the warming in the late twentieth century to be something new, rising faster and higher than has been seen this millennium. Until about 1800, internal climate system variability and natural forcings such as volcanic eruptions and solar variability are able to account for the reconstructed climate variations. In contrast, natural causes are unable to explain the recent warming; no explanation seems to fit the bill except the rise in greenhouse gas concentrations.

Is the globe warming?

How shall we assess the warnings of climate change? We can test the forecast by testing the climate models. Can the same models predict past climate changes? Even better, can they make predictions that can afterward be tested?

We will begin with the question of how much warming has actually occurred. Answering this question is trickier than one might have thought because temperatures vary from place to place and through time, and measurement techniques have changed over time. One measure of success is to find global temperature reconstructions from independent methods that agree with each other.

The next task is to come up with the histories of natural and anthropogenic ***climate forcings*** (Fig. 10.2). A climate forcing is a change in some driver for the temperature of the Earth. The big ones are solar variability, dust and aerosols in the atmosphere, and changes in greenhouse gas concentrations. Humans affect climate most strongly through greenhouse gas concentrations and aerosols, plus there are a few other, smaller, so-called ***anthropogenic climate forcings***. Some factors that affect climate are considered internal to the climate system, like the water vapor feedback, clouds, and changes in albedo. These are part of the model, not considered forcings but rather feedbacks.

We can compare different climate forcings in terms of their global average ***radiative impact***, their effect on the energy budget of the Earth, in units of watts per square meter. One watt per square meter of global average radiative impact from one climate forcing is broadly equivalent to 1 W/m^2 from a different climate forcing. How large are anthropogenic climate forcings relative to natural climate forcings? How well are these radiative impacts known?

The third step is to use the models to predict the past, using the various forcings. A prediction of the past is called a ***hindcast***. Can the model predict the past, using all of the climate forcings we know of? Within the uncertainties of the natural climate forcings, is it possible to explain the recent temperature record *without* anthropogenic climate forcings? This is a question that gets asked, implicitly, whenever it is claimed that the observed warming can be explained by the Sun getting brighter, for example. Is there any reason why the climate should be more responsive to 1 W/m^2 of forcing from one source and not responsive to a similar 1 W/m^2 from another source? But say that it were, say that anthropogenic climate forcings such as rising CO_2, for some unknown reason, do not lead to global warming. Is it possible to explain the temperature record without them, within all the uncertainties?

The thermometer records

The warming signal we are looking for is small compared with the variations in temperature from day to day, season to season. We want to know whether the planet surface as a whole is getting warmer, so we will have to average out all the weather. It could be treacherously easy for some bias in the raw data to be undetected or otherwise not corrected for. A computed average temperature for some particular place has to be balanced between daytime and nighttime temperature measurements, for example, and summer versus winter. If we want to compare the average temperature between one year and another, we also have to worry about the possibility that the way temperature is measured might have changed with time.

On land there is a potential source of bias known as the ***urban heat island effect.*** Sunlight hitting pavement is converted into warmer temperatures, what climatologists call ***sensible heat***, "sensible" because you can measure it with a thermometer. Some of the sunlight hitting vegetation drives water to evaporate and is carried away as ***latent heat*** (Chapter 5), which you can't feel or measure with a thermometer until the water condenses. A vegetated landscape doesn't warm up as much as a paved landscape because some of the heat is escaping invisibly, or insensibly I suppose you would say, as water vapor. This is an observed phenomenon; cities can be 5°C warmer than their surrounding countryside. Nighttimes in particular are warmer in cities. What if we established a weather station in some idyllic rural area, only to have the landscape urbanize over the decades? The temperature record would show a warming, which would be real for that location, but will confuse you if you try to infer the global average temperature trend from it. The warming in the city is real and should be counted in the global average, but if the data were biased toward cities, the computed average temperature could be higher than the real average temperature of the Earth.

Four independent studies of the land temperature data made an attempt to throw out data from locations that might have seen urbanization (Fig. 11.1). The studies all agree that removing urban locations has only a small impact on the global average temperature trend that you calculate from the data. They find that it doesn't matter how you deal with the urban data, you get pretty much the same answer.

The sea surface temperature, known to seagoing types as ***SST***, used to be routinely measured by heaving a bucket into the ocean to collect surface water and sticking a

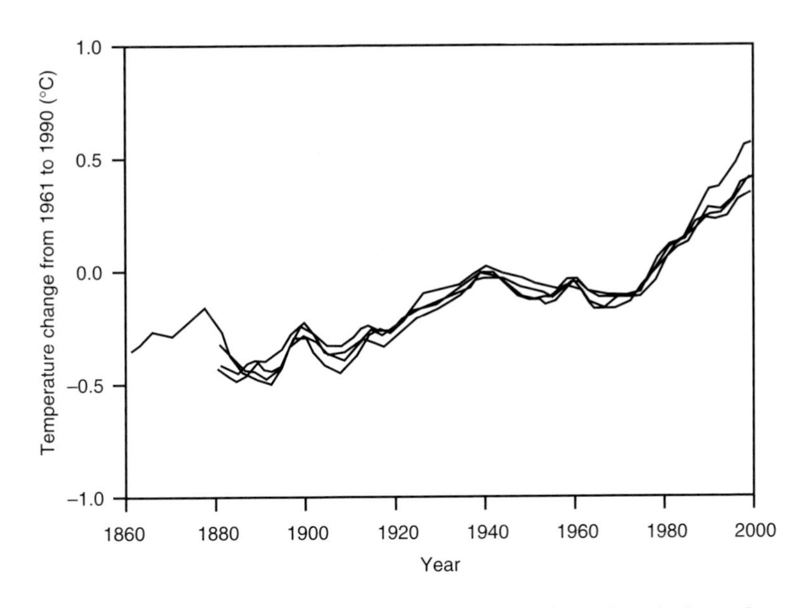

Fig. 11.1 **Instrumental land temperature reconstruction from four independent studies. Replotted from IPCC (2001).**

thermometer into the bucket on deck. Some of the buckets were made of wood, others of canvas. Evaporation of water stole some heat from the surface of the bucket, cooling the water somewhat. Canvas buckets cooled more than wooden ones because wet canvas is a poor insulator. After 1942, mariners began measuring the temperature of surface water as it was pumped into the ship for engine cooling water. This is a more accurate measurement but there is still the possibility of bias. The waters in the surface ocean sometimes get cooler with depth. A large ship might sample subsurface waters, which would be systematically cooler than the real SST. These effects must be corrected for, if we are to calculate a global average value of SST through time.

The land and sea records are consistent with each other, in that the SST record can be used to drive an atmospheric climate model, and the temperature history of the atmospheric model on land matches the reconstructed land surface temperature record (Fig. 11.2). You can see that the largest correction that had to be applied to the temperature records was the bucket correction for the SSTs. The abrupt change in SST in 1942 resulted from a switch between bucket and engine room temperature measurement.

Michael Crichton in his fiction book *State of Fear* shows temperature records from selected locations around the world, where temperatures have dropped over the past few decades. Climate has regional variations, this is true. The global average temperature rise incorporates these regions of cooling, but still shows warming on average. There are other places that Crichton does not show, the Arctic for example, where temperature has risen far more than the global average. The global average temperature record is based on thousands of individual locations around the world. The existence of a few where climate is cooling does not disprove the warming trend in the global average.

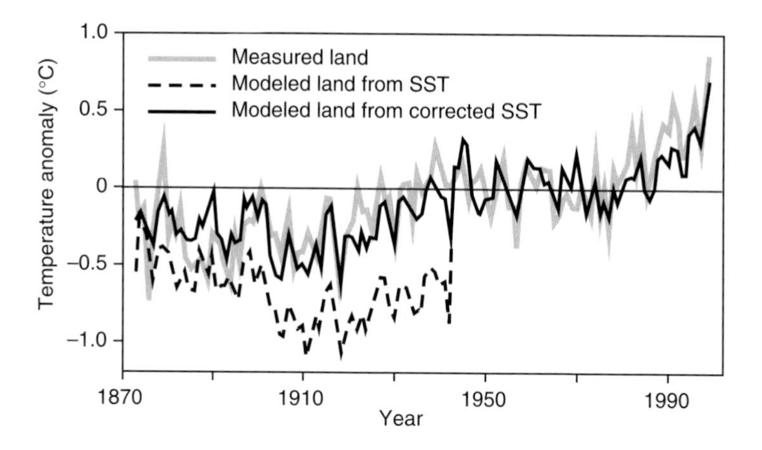

Fig. 11.2 **Land temperatures reconstructed from land thermometers and predicted using SST reconstructions, corrected and uncorrected for the bucket effect. Replotted from IPCC (2001).**

Temperatures measured from satellites

Satellites provide another source of global temperature trend information by measuring microwave light emission from oxygen gas, O_2, in the atmosphere. Because of the way that electrons are distributed around the nuclei of oxygen molecules, the rotation of oxygen molecules emits and absorbs energy in the microwave range of light. This is longer wavelength, lower energy light than the IR light that greenhouse gases absorb and emit. Microwave light does not carry much energy in the Earth's energy budget, so we have not been considering O_2 as a greenhouse gas the way we do CO_2. The difference is that microwave emission comes from rotation rather than vibration, and the frequency of light this generates falls outside the main blackbody spectrum of the Earth (Chapter 4).

The intensity of microwave radiation emitted from O_2 increases with increasing temperature, just as the intensity of IR light goes up with temperature. The satellites measure microwave emission in a range of wavelengths, and from these measurements construct temperature estimates for several regions of the atmosphere (see http://mtp.jpl.nasa.gov/intro/intro.html). The lowermost region of the atmosphere from which the satellite estimates the temperature spans from the surface to about 8 km altitude, with an exponential decay, rather similar in shape to the curve describing pressure with altitude (Fig. 5.2).

These satellite temperature estimates have been the subject of considerable scientific and political discussion because for many years the satellite estimates of warming disagreed with the reconstructions from thermometers. This has now changed, as scientists learn how to calculate temperature from the raw satellite sensor data (solid black line in Fig. 11.3). The satellite temperature record comes from a series of satellites which must be well calibrated against each other as they are pieced together into a longer composite record. The raw data must be corrected for things like changes in the orbit of the satellite. The current calibration of the satellite data shows good agreement with the land and sea instrumental records.

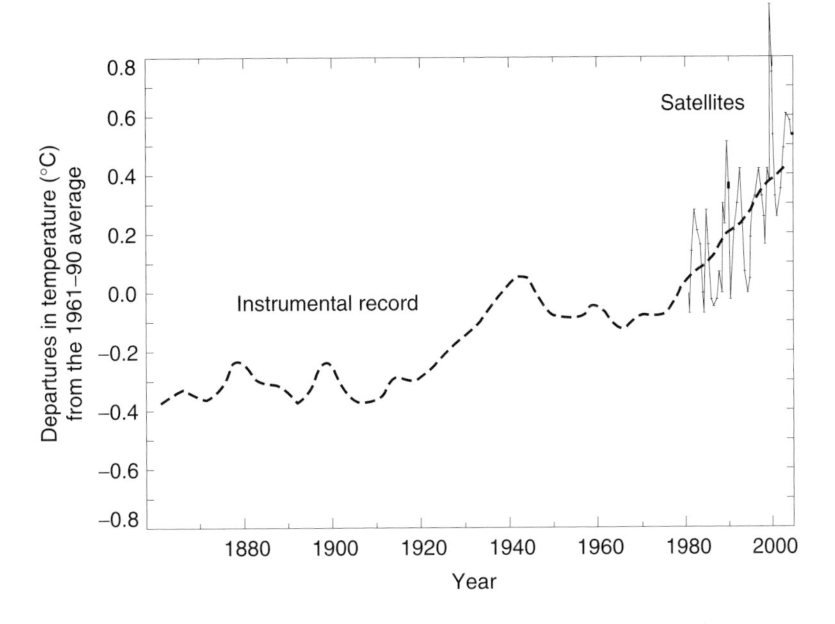

Fig. 11.3 **A comparison of the global average temperature reconstruction with the satellite record. Data replotted from Mears and Wentz (2005).**

Fig. 11.4 **The melting of the Qori Kalis glacier in the Peruvian Andes.**

Glaciers

The most visual indication of climate change is the melting of mountain glaciers. The vast majority of the mountain glaciers of the world are melting back at astonishing rates (Fig. 11.4). This is happening all over the world. Figure 11.5 shows records of shortening of 20 glaciers around the world. Many of the glaciers have been melting back since the 1700s as a result of natural warming since the Little Ice Age. However, melting has accelerated in the past decades. The Snows of Kilimanjaro (see Further reading) are forecast to melt entirely by 2030, as are all the glaciers in Glacier National Park in the United States.

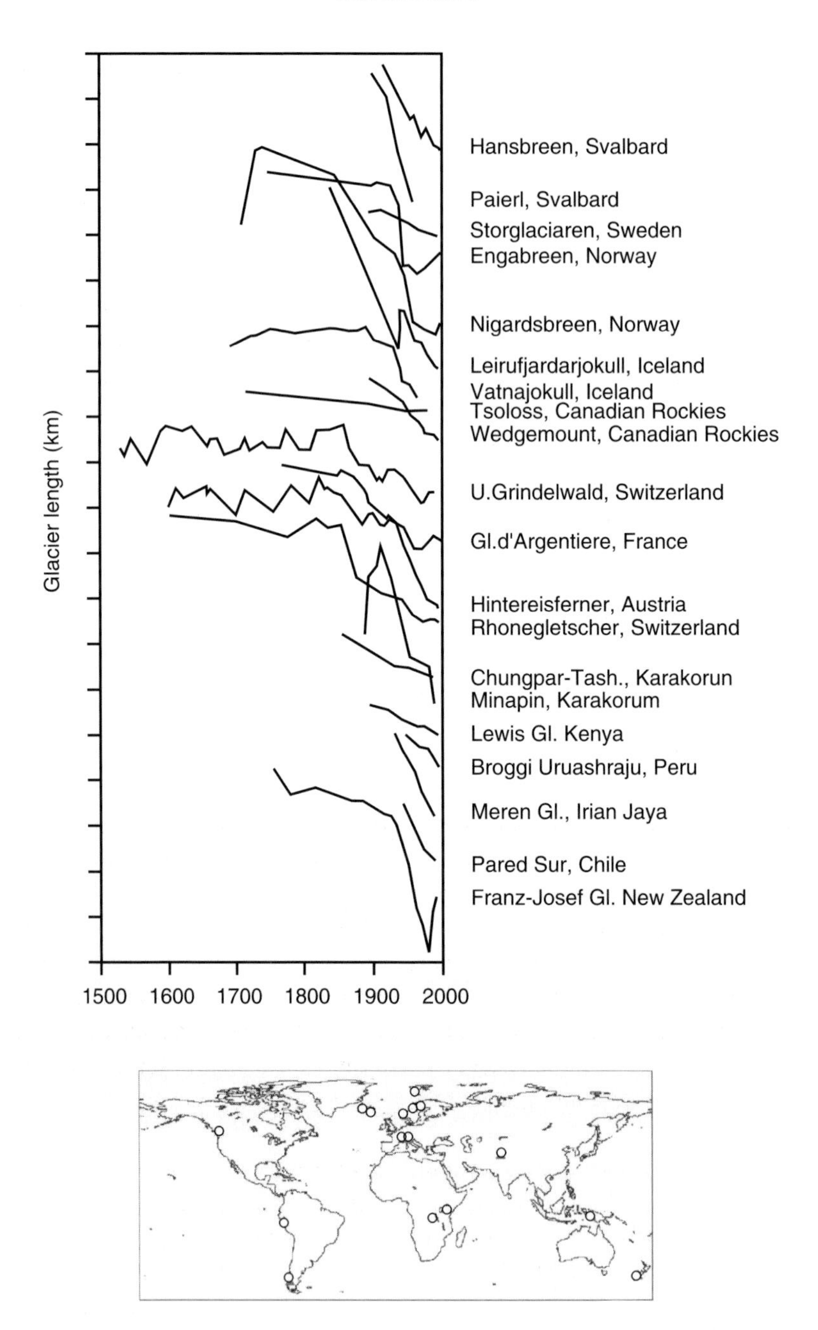

Fig. 11.5 **Glaciers are melting all over the world. Replotted from IPCC (2001).**

Climate forcings in the past

The next step is to ask about natural processes that drive the climate, natural climate ***forcings***. Different climate forcings can be compared with each other in terms of their energy impact on the Earth in watts per square meter. One watt per square meter in solar output, for example, is roughly equivalent to 1 W/m^2 of IR light trapped by CO_2, or at least these have about the same impact on model climates.

We can roughly estimate the climate impact of radiative forcings using another version of the ***climate sensitivity***. We have already discussed the climate sensitivity to doubling CO_2, called ΔT_{2x}. A middle-of-the-road estimate for ΔT_{2x} is 3°C. We can express the climate sensitivity as temperature change per watts per square meter of heat forcing. A typical estimate $\Delta T/(W/m^2)$ would be $0.75°/(W/m^2)$. The current anthropogenic climate forcing today is about 2.5 W/m^2 (Fig. 10.10). Multiplying by the climate sensitivity gives us a temperature change estimate of about 1.8 K. This is larger than the Earth has warmed because it takes time for the climate of the Earth to reache equilibrium (Chapter 12).

Solar intensity varies over the 11-year sunspot cycle by about 0.2–0.5 W/m^2 (Fig. 11.6). Solar intensity in the past is estimated by the accumulation rate of isotopes that are produced by cosmic rays, such as ^{10}Be and ^{14}C. These are called ***cosmogenic isotopes***. The idea is that a brighter Sun is a better shield to cosmic rays. Cosmic rays produce cosmogenic isotopes. So a high deposition rate of ^{10}Be at some depth in an ice core tells us that the Sun was weaker, unable to deflect cosmic rays, and so we infer that it was not very bright at that time. Who made that story up, you may wonder. We do in fact observe a correlation between cosmic rays, ^{10}Be production, and solar intensity over the 11-year sunspot cycle in recent times (Fig. 11.7). There are times in the past such as the Maunder Minimum, from about 1650 to 1700, when there were no sunspots. This was the coldest period in Europe in the last 1000 years, and glaciers advanced all over the world, including in the southern hemisphere, so it's pretty clear that the solar luminosity was less then than it is now. However, we have no direct solar measurements from a time when the Sun acts like this, so we don't have a very solid constraint on how much lower the solar output was. You can see that there is a factor-of-two uncertainty in the solar forcing in the past, as the thickness of the gray region in Fig. 11.6. The time history of solar forcing variability is to drift up and down on timescales of about 100 years.

Volcanic eruptions inject particles into the atmosphere, sulfate aerosols and dust. Particles injected into the troposphere last a few weeks before they are rained out. It doesn't rain in the stratosphere, so particles spend several years floating around before they eventually sink out. The Mt. Pinatubo eruption in 1991 cooled the planet by 0.5°C for several years, a result consistent with a middle-of-the-road climate sensitivity ΔT_{2x} of about 3°C, by the way. ***Volcanic climate forcings*** look like randomly spaced spikes of random but high intensity in Fig. 11.7.

To the two natural climate forcings we add the two main anthropogenic ones, ***greenhouse gases*** and anthropogenic ***aerosols***. Neither was important before about 1750. Greenhouse gases warm whereas sulfate aerosols cool. The greenhouse gas radiative impact is fairly uniform globally whereas the climate forcing from sulfate aerosols is concentrated in the northern hemisphere, downwind from industrialized and heavily populated areas.

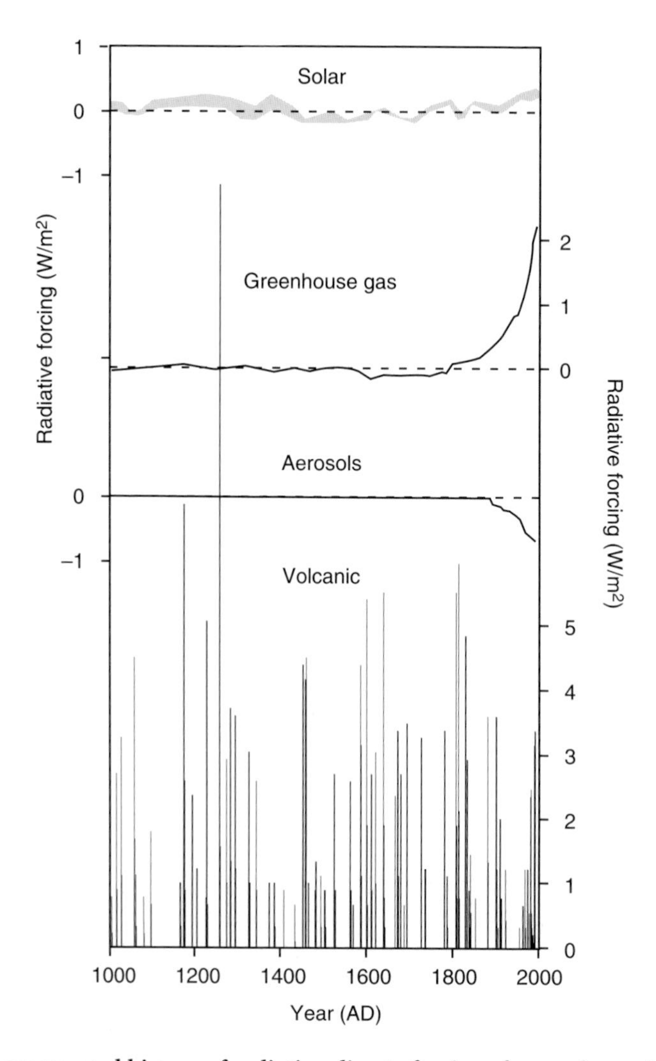

Fig. 11.6 **Reconstructed history of radiative climate forcings from solar variability, greenhouse gases, anthropogenic aerosols, and volcanic particle emission to the stratosphere. Replotted from Crowley (2000).**

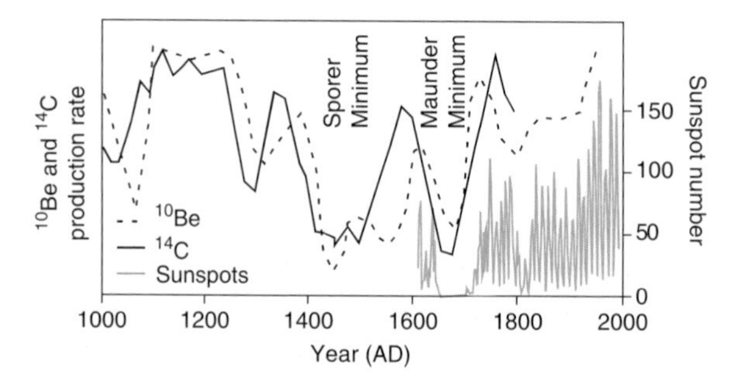

Fig. 11.7 **History of sunspot number and cosmogenic isotope production over the past 1000 years. Data from Beer (2000) and Lean (2000).**

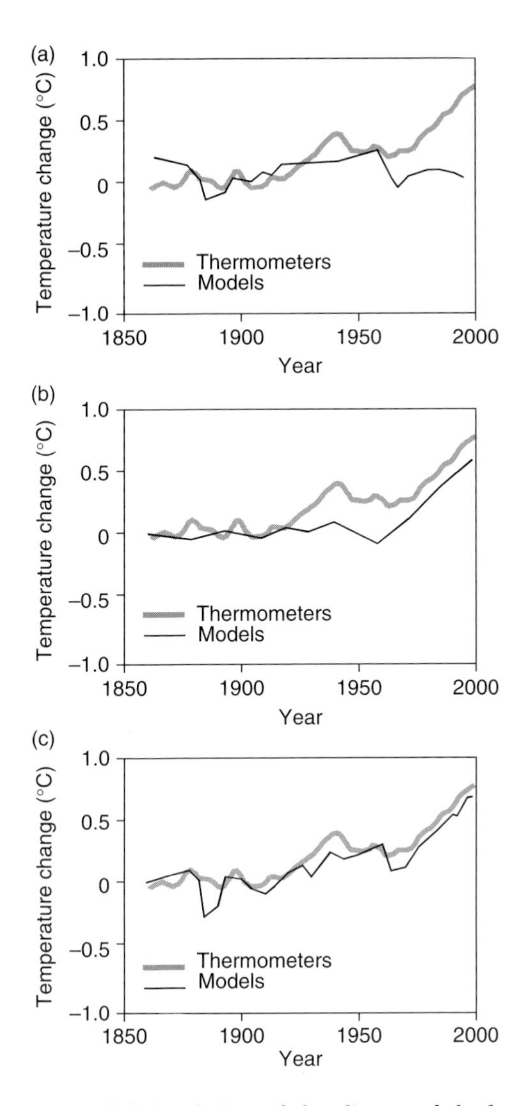

Fig. 11.8 **Hadley Centre model simulation of the climate of the last 140 years using (a) natural forcings only, (b) anthropogenic forcings only, and (c) all forcings. Replotted from IPCC (2001).**

Model data comparison

A comparison of the historic temperature record with model results is shown in Fig. 11.8. The observed temperature rose in the period from 1900 to 1950, remained steady or even declined from 1950 to 1970, and then started rising again from 1970 to today. The model temperature (from the British Hadley Centre model, one of the most advanced climate models) is shown for three different model simulations. One uses only natural climate forcings: solar and volcanic dust. This simulation captures the temperature rise in the early part of the twentieth century, but does not reproduce the warming in the last few decades. Another simulation uses only the anthropogenic climate forcings. This captures the present-day warming but not the warming in the early part of the twentieth century. Only when natural and anthropogenic forcings are combined in a model simulation does the model do a reasonable job of simulating the observed temperature changes. A reasonable conclusion would be that the global warming forecast model seems to have the sensitivity to greenhouse gases about right.

The IPCC 1995 report predicted that the warming trend already apparent at that time would continue because of rising greenhouse gas concentration, and they were right. The year 1995 was the warmest on record at the time, but that record has been matched or broken by 6 of the 9 years since then (Fig. 11.3).

The climate of the last millennium

We would like to extend the temperature record further back in time, before the era of technological measurements. The thermometer and satellite temperature records indicate a warming trend since about 1970, but this begs the question: is this warming normal, or is it something new? Temperatures from before the days of thermometers can be estimated by a variety of tricks, measurements of something we can measure, interpreted as indicators of things we would like to know. These alternative measurements are known as ***proxy*** measurements.

One source of proxy temperature measurements is ***tree ring*** thickness (Fig. 11.9). In places where winter is much colder than summer, wood that grows in winter is a different color than wood from summer, leading to the formation of annual rings. Trees

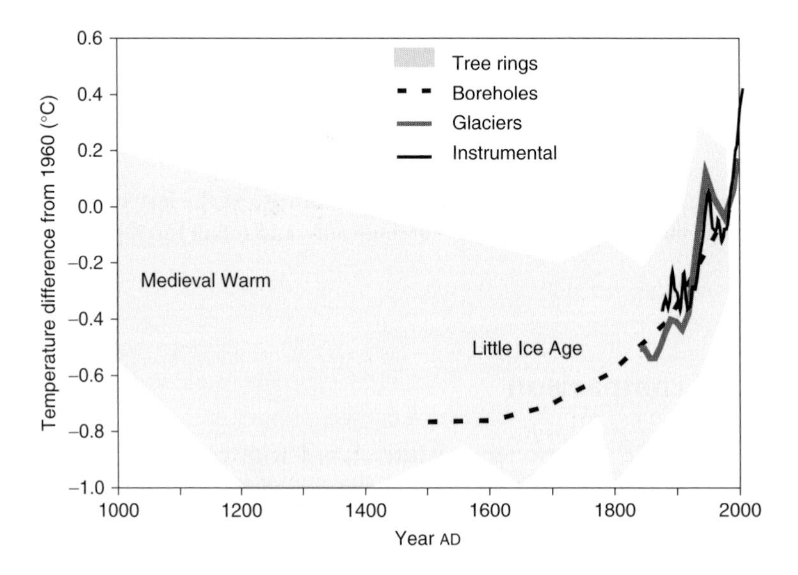

Fig. 11.9 **Northern hemisphere mean temperature reconstructions. Tree rings from IPCC (2001), boreholes from Rutherford and Mann (2003), and glacier length from Oerlemanns (2005).**

grow more quickly during warmer years than colder years, so it makes sense that tree ring thickness might carry information about temperature. Long records of tree ring thicknesses can be assembled from collections of many individual trees that lived long ago, by matching the patterns of widths to fit the different trees together in time. The widths of the rings must then be ***calibrated*** against temperature changes from part of the record where the temperature is known independently. This is tricky because trees grow or don't grow in response to other environmental variables, such as the availability of water, nutrients, and even atmospheric CO_2. Young trees grow more vigorously than old ones. Temperatures vary from place to place, and so calculating a global average from local data introduces another possibility of bias.

Other proxy measurements for temperature include the temperature as a function of depth in the Earth and in ice cores, called ***borehole temperatures***. The temperature goes up with depth in the Earth or in the ice because heat is being conducted from the interior of the Earth to the surface. The steeper the temperature change with depth, the more heat is transported. If we imagine holding the temperature at the Earth's surface constant for a long time, then the temperature profile with depth in the Earth or the ice would be a straight, linear increase with depth (Fig. 11.10). If we now imagine warming the surface of the Earth, then the warming will propagate down into the Earth or the ice. If we measure the temperature profile today, we can use mathematical methods called inverse methods to figure out what the surface temperature was in the past, in order to generate what we see today. Borehole temperature records lose the little wiggles in the real temperature record, and memory fades with time so the uncertainty gets greater. Borehole temperatures show warming throughout the past 500 years, more or less consistent with the tree ring record (Fig. 11.9).

Other sources of information about past climates come from ice sheets and mountain glaciers. Glacier lengths have recently been used to construct a third independent

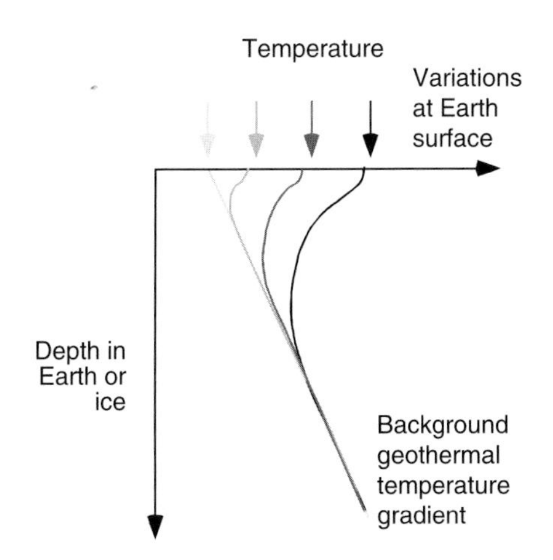

Fig. 11.10 **Time evolution of subsurface temperature, when the surface warms up. From a snapshot of the current temperature profile (black line), scientists can reconstruct the history of the surface warming (arrows).**

prehistoric temperature reconstruction in Fig. 11.9, in substantial agreement with the results from tree rings, boreholes, and where it exists, the instrumental temperature record.

Temperature reconstructions of the last millennium sometimes show a period of warmth from 800 to 1200, called the ***Medieval Warm*** period, and cooler temperatures from 1500 to 1800, called the ***Little Ice Age***. Tree ring records mostly come from the northern hemisphere, so it is not clear whether the Little Ice Age and Medieval Warm periods were global or not. The timings of warmings and coolings, where they are documented in the northern hemisphere, do not seem to be precisely synchronous.

Models are able to simulate the Medieval Warm and Little Ice Age, largely as a result of long-term changes in solar forcing. The little ice age in particular coincided with a time of low solar flux inferred from cosmogenic isotope production (Fig. 10.6), and the Maunder Minimum dearth of sunspots (Fig. 10.7). The climate record of the past millennium is a less stringent test of the climate models than the record from the last century, though, because there is more uncertainty in both the radiative forcing and in the temperature record we are trying to match. However, we can draw the striking conclusion from the millennial climate record that the warming of the twentieth century, especially in the last few decades, stands out as something new that has not happened before. The warming we observe in the past decades is more intense, faster, and more obviously global than the Medieval Warm or Little Ice Age periods in the past.

Climate variations on orbital timescales

The glacial cycles are driven by variations in Earth's orbit. Climate reconstructions on these timescales come from ice cores and ocean sediment cores. We looked at an ice core data set in Fig. 8.3 showing concentrations of CO_2 and CH_4 in the atmosphere, recorded in bubbles in the ice.

Figure 8.3c, which we haven't talked about much yet, is a reconstruction of the temperature where the ice accumulated, in Antarctica (not a global average). The temperature estimate is based on the relative abundances of different ***isotopes*** of hydrogen and oxygen in the water. Two isotopes of an element have the same number of electrons and protons, and hence pretty much the same chemistry. That is what we mean when we say that they are both the same element. The isotopic sisters have different number of neutrons, however, so they have different masses. Some isotopes, like carbon-14, decay radioactively. Other isotopes, such as oxygen-16 and oxygen-18, are both perfectly stable and do not decay.

The different masses cause two isotopes of the same element to behave slightly differently from each other in chemical reactions. These are subtle effects; the raindrops might have a ratio of oxygen-18/oxygen-16 that is only a percent or so different from the isotopic ratio of the vapor. In order to make the numbers a little easier to remember, isotope geochemists describe ratios not as percentage differences but rather as ***per mille***, written as ‰. One percent is one part in 100, whereas one per mille is one part in 1000. Ten ‰ = 1%.

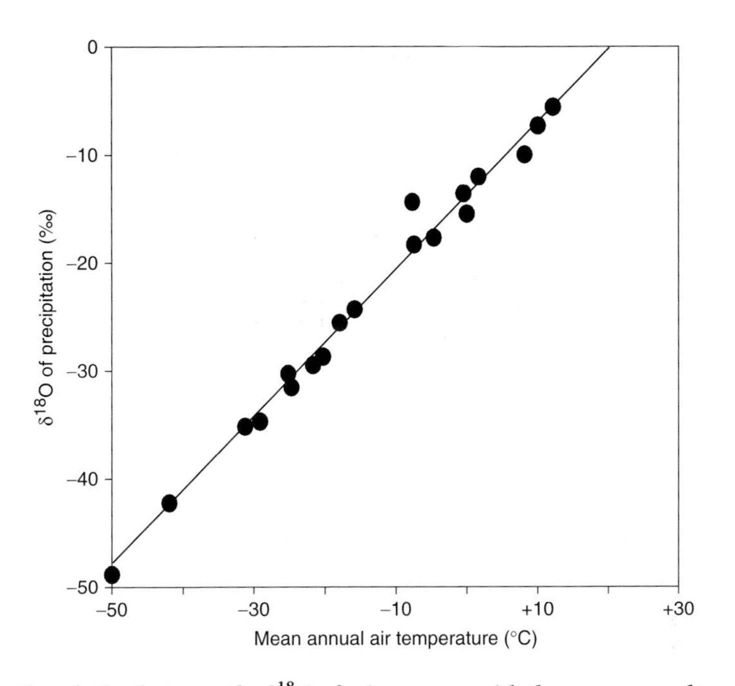

Fig. 11.11 **Correlation between the δ^{18}O of rain or snow with the mean annual temperature where the precipitation fell.**

One example of a process that affects isotopic ratios is rain and evaporation. A water molecule comprised of a heavier isotope of oxygen or hydrogen will tend to rain out earlier than would a water molecule made of the lighter isotope. As the temperature over an ice field decreases, the water vapor that remains in the atmosphere to snow down on the ice becomes increasingly depleted in the heavy isotopes. For this reason, there is a systematic relationship today between the temperature and the relative abundances of the heavy and light isotopes in water. Snowfall has less heavy isotopes left when the temperature of snowfall drops (Fig. 11.11). A small caveat: the data points in Fig. 11.11 are all present-day precipitation. If the oxygen isotopic composition of the evaporating source water were different in the past, the entire oxygen-isotope/temperature line might move up and down. Scientists use a combination of oxygen and hydrogen isotopes to correct for this effect.

Sediment cores also provide information about past temperatures. The oxygen isotopic composition of $CaCO_3$ shells tells us something about past temperatures, as do measurements of the chemical composition of the shell. A family of protozoa called the ***foraminifera*** supply most of the shells that are analyzed from deep sea cores. Some, like the ***planktonic*** foraminifera, live in the sunlit surface waters and leave a record of SSTs. Others, the ***benthic*** foraminifera, live on the sea floor, and can tell us about past deep ocean temperatures.

The surface of the Earth was perhaps 5–7°C cooler during the ***Last Glacial Maximum***, about 20,000 years ago (Fig. 8.3). The changes in greenhouse gas concentrations changed the radiative forcing of the Earth, by about 2–3 W/m^2. There was an even

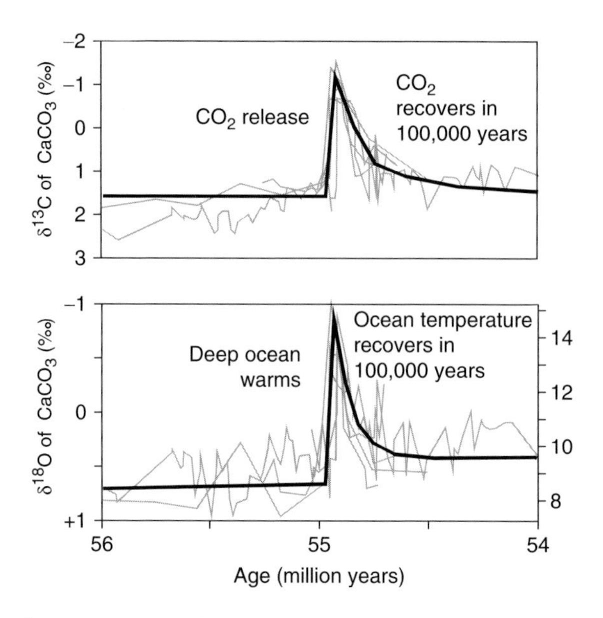

Fig. 11.12 **The Paleocene Eocene Thermal Maximum event. Replotted from Zachos (2001).**

larger change in the radiative forcing from albedo changes, perhaps 4–7 W/m^2. This is the result of ice sheets, sea ice, and changes in the terrestrial biosphere. Here is another chance to test the climate models. There are uncertainties about the circulation of the ocean during glacial time, but in general, climate models are able to simulate the observed cooling of the Earth's surface. If the climate models were somehow too sensitive to greenhouse gases, then we would expect the climate models to predict a colder Last Glacial Maximum than we piece together from the proxy records. CO_2 is only part of the reason for the colder glacial climate, but it is striking to look at how similar the CO_2 and temperature records are. Every little wiggle lines up between the two records. It is difficult to argue that CO_2 is somehow unimportant to climate, after looking at this record (Fig. 8.3).

The Paleocene Eocene Thermal Maximum

A final test of the global warming forecast comes from an interval of time 55 million years ago called the Paleocene Eocene Thermal Maximum event or ***PETM***. This is much further in the past than any time we have considered yet, and the data are somewhat scantier.

We have evidence from the isotopes of carbon that a sizable amount of carbon was released to the atmosphere or ocean (Fig. 11.12). The carbon that was released had a different ratio of carbon-12 to carbon-13 than the CO_2 in the atmosphere and ocean, so it changed the isotopic ratio of the ocean and therefore also the ratio recorded in $CaCO_3$ formed in the ocean. It is difficult to know how much carbon was released because we don't know its carbon isotopic ratio. If it were methane, which has an extremely low ratio (negative $\delta^{13}C$), we would need a few thousand gigatons carbon. If it were organic carbon, which has a less extreme $\delta^{13}C$, a larger slug would be required to explain the $\delta^{13}C$ spike. For comparison, the entire inventory of minable coal, using present-day technology, is about 5000 Gton C (Chapter 8). The PETM is therefore an analog to the potential effect of fossil fuel combustion.

The isotopes of oxygen provide an indication of the warming of the deep sea that resulted from the carbon release. Temperatures were already somewhat warmer in the deep ocean than they are today before the carbon release. After the carbon was released, temperatures warmed by about 5°C.

It is impossible to use the extent of warming as an indication of the climate sensitivity of the Earth at that time because we don't know what the CO_2 concentration in the atmosphere was. However, we can use this data as a test of the prediction, from Chapter 10 and Fig. 10.3 that it will take hundreds of thousands of years for temperatures to recover to their natural baselines, after a sizable fossil fuel CO_2 release.

Take-home points

1. The past few years have been warmer than the thermometer records of the past 140 years, and warmer than reconstructed records over the past 1000 years.
2. Climate varies naturally on all different timescales, even under steady forcing. Models capture this natural variability, more or less, and lead us to conclude that the recent warming is unlikely to be caused by natural variability.
3. Climate is also driven by natural and human-induced changes in radiative forcing, including volcanic eruptions, solar variability, greenhouse gases, and human-released aerosols. Models cannot explain the recent warming without including anthropogenic forcing.

Editorial: consensus in science

All three studies of land temperature records agree; what does that mean? You can imagine there might be a lot of judgment calls trying to eliminate urbanization from the huge morass of all the weather measurements made in the last hundred years. There is always the possibility of prejudice on the part of the researcher, who may allow his or her judgment on a myriad of minor issues to be swayed by an expected result.

Skeptics accuse the scientific establishment of encouraging the creation of false spurious consensus, by making it easier for the scientist who reinforces the consensus view to get grants funded and papers published. I would argue that, in my experience, the way a scientist becomes famous (and let's face it, we're not doing science for the money) is to successfully challenge the consensus view. Shoot the moon. The person writing the second paper on some topic would really like to find something that the first paper missed or maybe got wrong. It could be easier to get a boring consensus paper published; that is true. Extraordinary claims require extraordinary evidence. But, on balance, my opinion is that the structure of the scientific establishment as it is practiced in university and government laboratories around the world encourages the challenger, rather than stifling dissent unproductively. For this reason I feel that in general, replicate scientific studies do bolster a scientific conclusion. Michael Crichton in his fiction work *State of Fear* argues the opposite. You will have to decide for yourself.

Projects

Point your web browser to http://understandingtheforecast.org/Projects/bala.html. This is a system to browse the output of a climate model run to the year 2300, from Bala et al. (2005). The browser allows us to explore the output of a coupled atmosphere, ocean, and land surface model. The atmosphere model comes from the National Center for Atmospheric Research in Boulder, Colorado, and the ocean model comes from Los Alamos National Laboratory in New Mexico.

The browser allows several options for making plots.

1. There are two model runs, one with CO_2 emissions simulating the eventual combustion of the entire coal reservoir, about 5000 Gton C. This run is labeled "***Rising*** CO_2" on the experiment pull-down. The other model run is called a ***Control***, in which there are no CO_2 emissions. The third is a "***Drift-corrected***" option that will subtract changes in the control, the model drift, from the rising CO_2 run.
2. Some of what you will see is simple multi-year variability. In order to remove this, perhaps for comparing one climate state with another, select the ***averaging, 10-year climatology*** option.
3. You can highlight the changes in a climate variable by subtracting one from another, to say things like "it is 2°C warmer than it used to be." This is called an ***anomaly*** plot.
4. Push the ***Plot*** button to generate a contour plot.
5. Wave the pointer over the map to display the numerical value from that location on the plot. In principal, it should be possible to guess what the value should be using the colors and the contours, but it is often helpful to see a number expressed as a number, instead of as a color.
6. Click the mouse over a point on the map to bring up another type of plot entirely, a plot of the year-to-year variations in the global average of the variable. The averaging option does not apply to this type of plot.

a. ***Temperature.*** Bring up an annual mean temperature map. Some parts of the world warm quickly, leveling off within a century or so, while other parts of the world continue warming until the model run stops in the year 2300. Can you figure out what the fast places have in common and how they differ from the slow places?
   The annual average temperature in Chicago in the year 2100 is comparable to the natural annual average temperature where?

b. ***Vegetation types.*** A color key for the biome types is shown in the text. Compare the changes in the Arctic with the changes in lower latitudes, where does the change happen first?

Further reading

Beer, J., W. Mende, and R. Stellmacher, The role of the sun in climate forcing, *Quaternary Science Review, 19*, 403–415, 2000.

Crowley, T.J., Causes of climate change over the past 1000 years, *Science, 289* (5477), 270–277, 2000.

IPCC Scientific Assessment, 2001, Cambridge University Press, or downloadable from http://www.grida.no/climate/ipcc_tar/. Technical Summary.

Lean, J.L., Short term, direct indices of solar variability, *Space Science Reviews, 94* (1–2), 39–51, 2000.

Mears, C.A., and F.J. Wentz, The effect of diurnal correction on satellite-derived lower tropospheric temperature, *Science, 309* (5740), 1548–1551, 2005.

Oerlemans, J., Extracting a climate signal from 169 glacier records, *Science, 308* (5722), 675–677, 2005.

realclimate.org is a commentary site on climate science by working climate scientists for the interested public and journalists.

Rutherford, S., and M.E. Mann, Optimal surface temperature reconstructions using terrestrial borehole data (vol 108, pg 4203, 2003), *Journal of Geophysical Research-Atmospheres, 109* (D11), 2004.

Zachos, J.C., M. Pagani, L. Sloan, E. Thomas, and K. Billups, Trends, rhythms, and aberrations in global climate 65 Ma to Present, *Science, 292*, 686–693, 2001.

12
The forecast

There are three sources of uncertainty in the forecast of global mean temperature in the year 2100 (an arbitrary but widely used benchmark year). One is the equilibrium climate sensitivity, ΔT_{2x}, of the real climate. Another is the amount of time lag in the real climate as the oceans slowly warm and as various feedbacks such as with water vapor come to equilibrium. The third is the amount of greenhouse gas that will be released. In general, the forecast is for 2–5°C warming by 2100 if mankind follows the IPCC Business-as-Usual (BAU) scenario. This would be considerably bigger than the ~1°C temperature variations through the last 1000 years, the biggest climate change since the 6°C temperature change at the end of the last glacial cycle.

Rainfall is expected to increase overall, in particular in equatorial and high-latitude regions. The subsidence regions at 30°N and S are expected to dry, as are continental interiors. Changes in rainfall are more difficult to forecast than changes in temperature, but may be more important to maintaining a stable food supply. There is the potential for long-term regional droughts, such as the 500-year drought that accompanied the Medieval Warm period in Europe. The intensity of tropical cyclones (hurricanes) is expected to increase. This is based on model predictions and meteorological observations.

The IPCC scenarios can be viewed as best-case scenarios, in that they are smooth with no unexpected surprises. The glacial climate was subjected to savage climate swings which we of the Holocene have largely been spared. It is difficult to predict whether greenhouse warming could push our climate into some rearrangement such as these abrupt climate changes in the past.

Temperature

We will begin with estimates of the climate sensitivity, ΔT_{2x}. The IPCC Third Assessment Report (2001) (IPCC as an organization is described in Chapter 13) reports that the mean of the climate sensitivity, ΔT_{2x}, of 15 climate models is 3.5°C. The ***standard deviation*** of the climate sensitivities, denoted by the Greek letter sigma (σ), is 0.9°C.

A standard deviation describes the variability or spread of a list of numbers, in this case model ΔT_{2x} values. If the values are distributed in a normal "bell" curve, then the meaning of the standard deviation would be that 63% of the ΔT_{2x} estimates should fall within 1σ of the mean (in this case ±0.9°C), and about 95% of the estimates would fall within 2σ (±1.8°C). The statistics imply that with 95% certainty, a new estimate of

ΔT_{2x}, such as from some new model, should probably fall between 1.9°C and 4.1°C. This range of model results in one measure of how well we can forecast the future.

Another approach to estimating how well we know ΔT_{2x} uses a bit more brute force. Climate models have knobs, numbers that are not known precisely which affect the behavior of the model. These numbers are called ***tunable parameters***. Cloud droplets, for example, do not form immediately when the relative humidity exceeds 100%. They may start forming at 110% relative humidity, or 120%. The best value of this parameter to use in a model is not known precisely. It varies from cloud to cloud, no doubt. Our model does not resolve all of the processes that would enable it to predict cloud droplet formation. The model must be told what value to use.

A brute force method for estimating the uncertainty in the model forecast is to run a climate model many times using a range of values of many different tunable parameters, varying multiple parameters at the same time. Climateprediction.net uses donated "screen saver" computer time to tackle this job. Each set of model parameters was run multiple times, with slightly varying initial conditions, to generate an ensemble (Chapter 7). Stainforth et al. (2005) analyze over 2000 model runs, over 100,000 years of model time. The URL for this project is www.climateprediction.net. Perhaps you would like to contribute to the effort yourself.

The distribution of model runs is shown in Fig. 12.1a. Most of the model combinations predicted a ΔT_{2x} of about 3.4°C. There is a long "tail" to the distribution (this is not a normal "bell" curve), with a few very high ΔT_{2x} estimates, ranging to 11°C. On the other side, there were very few parameter settings that came up with ΔT_{2x} of much less than 2°C.

Each run included a control period, where CO_2 concentration remains constant, and a period with doubled CO_2. The climate of the control period was compared with meteorological data, and all the misfits added up into a single number. The misfit numbers from the runs are compared with the errors of a hand-tuned ("standard") model run. The misfits from the low-ΔT_{2x} models tended to be a bit higher than the error from high-ΔT_{2x} models (Fig. 12.1b) but none of the errors was bad enough to declare any of the models to be obviously wrong.

Uncertainty in ΔT_{2x} is only the beginning of the uncertainty in the temperature forecast for the coming century. ΔT_{2x} gauges the equilibrium response to doubling CO_2, but it takes quite some time for the climate system to reach equilibrium. Predicting the temperature in the year 2100 requires modeling what is called the ***transient response.*** IPCC uses a standard benchmark for comparing the transient responses of models, which they call Transient Climate Response (***TCR***) and define as the model temperature at doubled CO_2 concentration when the CO_2 has been rising at a rate of 1% per year (Fig. 12.2). Real atmospheric pCO_2 itself is not rising at 1% per year, but the idea is to raise CO_2 a bit faster, to account for the greenhouse forcing from methane and other greenhouse gases. If the radiative forcing in the future doesn't go up as fast as 1% per year, then the TCR would be more appropriate to compare with the radiation = doubled CO_2 year, whatever year that turns out to be. The average TCR values from 20 models is 1.8°C with a standard deviation of 0.4°C.

Climate takes a long time to change, that is to say, it has a long transient response to the change, in part because the ocean stores a lot of heat. During this time when we are

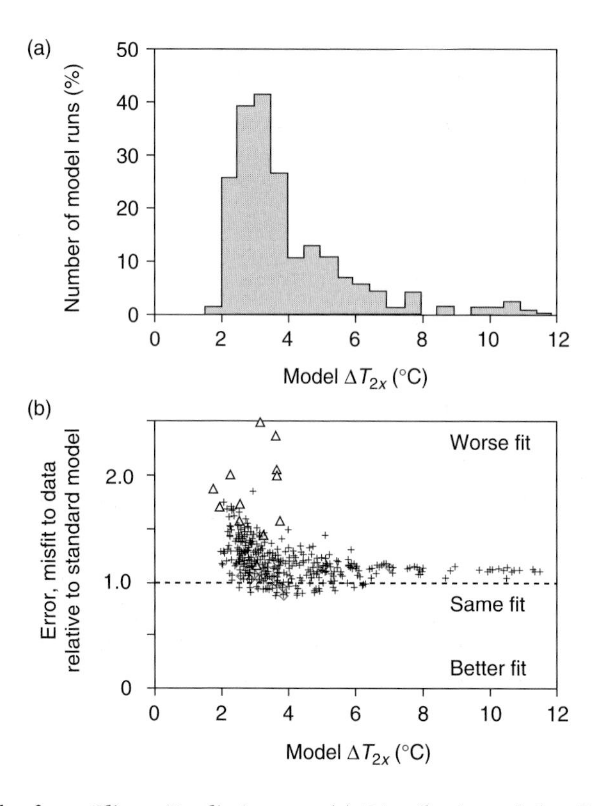

Fig. 12.1 **Results from ClimatePrediction.net. (a) Distribution of the climate sensitivity, ΔT_{2x}, of different model configurations. (b) Error in predicting the real climate, relative to the error of a hand-tuned model = 1.0.**

waiting for the Earth to warm up, there is an imbalance in the energy budget of the Earth. First a rise in greenhouse gas concentration decreases the outgoing flux of heat, and then eventually the Earth warms up, ramping up its outgoing energy flux, until the Earth's energy budget finds its new balance.

The energy budget of the Earth today, based on satellite measurements, is out of balance by about 0.75 W/m^2. The excess heat is being absorbed into the ocean. We can measure that the temperature is rising in the upper ocean, and calculate that the ocean is where most of the 0.75 W/m^2 of heat is going. As the ocean warms, it allows the climate of the Earth to warm, increasing the energy flux back to space, until the energy budget of the Earth reapproaches a state of balance (Fig. 12.3). It will take centuries to warm up the oceans to a new equilibrium temperature distribution. The oceans are keeping us cool.

The warming from rising greenhouse forcing takes a long time also because of feedbacks in the climate system, such as the ice albedo and the water vapor feedbacks, discussed in Chapter 7. The effect of these feedbacks is to make it harder for the poor old Earth to balance her energy budget. I won't guarantee it, but it might be helpful to look at Fig. 12.3 again. An increase in temperature of the Earth increases the water vapor concentration, a greenhouse gas that blocks infrared energy loss to space.

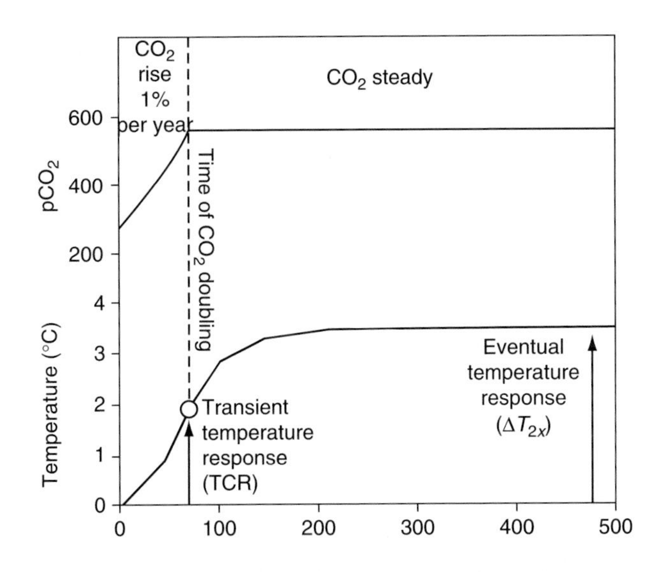

Fig. 12.2 **TCR is a diagnostic for comparing one model with another. It is a snapshot of the temperature at the time when CO_2 reaches double the preanthropogenic value. If CO_2 is held constant at that level, the climate continues to rise, until, in equilibrium, the temperature reaches the climate sensitivity, ΔT_{2x}.**

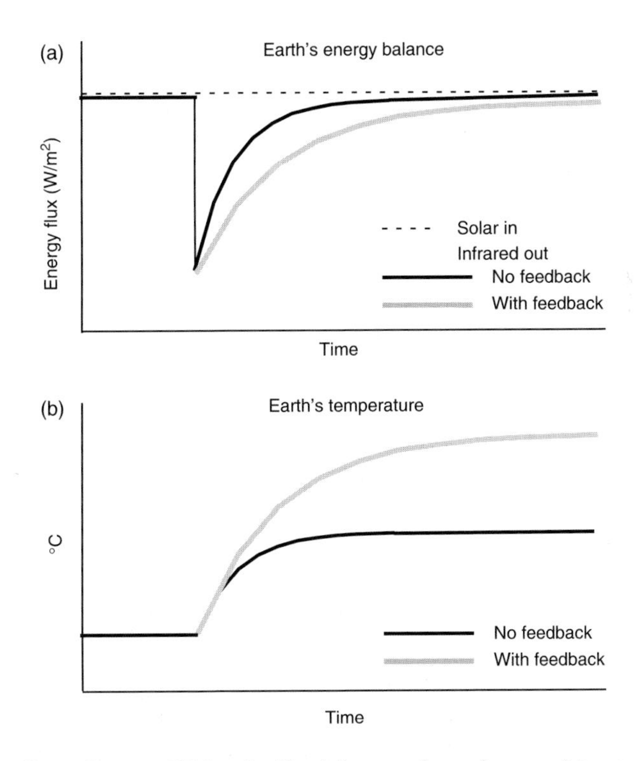

Fig. 12.3 **The effect of an amplifying feedback is to prolong the transition to a new climate, as well as increase the final temperature change. (a) Earth's energy balance and (b) Earth's temperature.**

The outgoing energy flux does not increase as much with rising temperature (gray solid line in Fig. 12.3) as it would if there were no feedbacks (black solid lines in Fig. 12.3). The existence of the feedback slows down the approach to the new climate equilibrium, as well as making the final temperature change larger.

So the amount of time it will take to balance the energy budget depends on two things. One is the heat uptake by the ocean, and the other is the strength of the feedbacks such as water vapor. One estimate of the equilibration time for climate is about 60 years. If the real feedbacks turn out to be stronger than we think, this will have two consequences. One is that the ultimate temperature increase will be greater, and the other is that it will take longer to reach the ultimate temperature increase. The best guess is that about 40% of the warming that will occur from the CO_2 already released, what is called ***committed warming***, has yet to take place. We have paid for 1°C warming, but we have so far received only 0.6°C.

Transient climate runs forced by the IPCC BAU scenario predict temperatures 2–5°C warmer by the year 2100 (Fig. 12.4). The uncertainty in our forecast for the temperature in the year 2100 derives from two sources, which contribute about equal to the uncertainty. One is the uncertainty in model temperature response to a given amount of CO_2, and the other is uncertainty in what our CO_2 emissions will be in the future.

A temperature change of 2–5°C may not sound like very much. The daily cycle of temperature is greater than that, to say nothing of the seasonal cycle of temperature. The main impacts of future climate change may come from changes in rainfall, rather than temperature. But the temperature change by itself is far more significant to the landscape of the world than you might think.

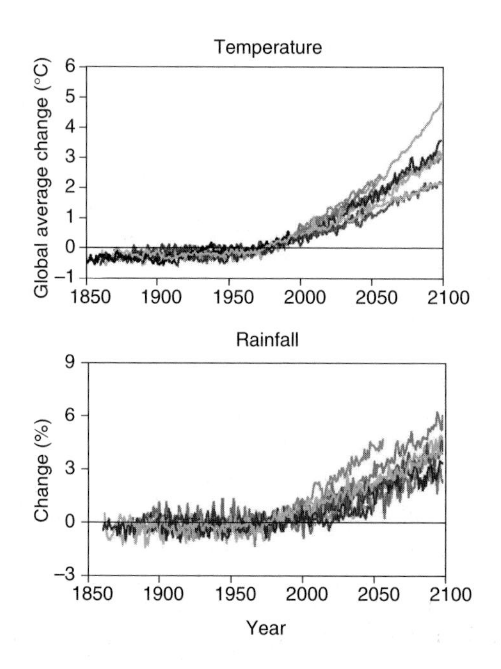

Fig. 12.4 **Model simulations of the BAU scenario, from IPCC 2001.**

One point of comparison is the temperature difference between now and the last ice age, estimated to be about 6°C. This was a huge climate change. If we were looking forward to a Glacial Maximum in the future, rather than backward into the safe past, we would be in a panic. It would be apocalyptic. Europe was an occasionally habitable tundra. All the vineyards and beer gardens, forget about them. The ice sheet in North America dwarfed what exists today in Greenland. Greenland, in spite of its name, is a pretty fierce place. Of course, Europe and North America were extreme cases because the ice sheets were here, but the landscape looked different around the globe. Changing snowlines make it clear that it was noticeably colder. Pollen data show huge changes in vegetation type. The coastlines were barely recognizable. Truly it was a different world.

Another comparison is to the Little Ice Age and Medieval Warm periods (Chapter 11). Some reconstructions of the global mean temperature, or Northern hemisphere temperature, show temperature changes of perhaps 0.5–1°C. These climate intervals were not the end of the world, but they definitely sufficed to rearrange civilizations. In medieval time, European agriculture was a bounty of plenty, in a stable, benign climate. Meanwhile a 500-year drought coincided with the demise of two organized civilizations in the New World, the Classic Maya and the Anasazi. The Little Ice Age climate was much more unstable than it was in medieval times. Temperature or rainfall would change suddenly for a year, or decades. There were periods of drought, periods of hot summers, of arctic winters, and of mild periods of moderate climate and good harvests. The years 1690–1730, roughly coincident with the Maunder Minimum, saw sea ice around Britain and northern France and a complete rearrangement of the fisheries in the Atlantic.

The impression I have is that a temperature change of 1°C is probably not a world-shattering change, at least globally, although there is the risk of essentially perpetual regional droughts, such as occurred in the American Southwest during the medieval warm time. By analogy to the intensity of the climate changes that came with the end of glacial time, I would guess that a global mean temperature change of 5°C would be catastrophic.

The distribution of the forecast temperature change is not uniform geographically or in time. Plate 12.1 shows the mean annual temperatures from a climate model described in the Projects section and in Bala et al. (2005). The atmosphere model was developed at the National Center for Atmospheric Research, a government agency in Boulder, Colorado. The ocean model was developed at the Los Alamos National Lab in New Mexico. In general, this particular model has a relatively low climate sensitivity, ΔT_{2x}, of 2–3°C for doubling CO_2. The temperatures are plotted as anomalies, differences from the temperatures in year 2000 in Plate 12.2.

The high latitudes warm more than low latitudes, by a factor of 3 or 4, because of the ice albedo feedback. Winter temperatures in Alaska and Western Canada have warmed by 3–4°C, compared with a global average of perhaps 0.5°C. Much of the high-latitude land surface is ***permafrost***, defined as soil that is frozen year round. The surface of the soil may melt in summer, a layer in the soil called the ***active zone***. As temperatures rise, the active zone gets thicker. As subsurface ice melts, the soil column collapses, leaving houses and trees tilted at crazy angles. Most of the Trans-Alaska oil pipeline has its foundation in permafrost soil. Lakes suddenly drain away, as melting ice leaves holes in the soil column. Coastlines are collapsing at rates of 40 m/year in some parts of the Canadian Arctic and Siberia. Models predict that the tundra ecosystem may disappear almost entirely in the coming centuries.

In mid-latitudes and the tropics, the day-to-day impact of the temperature change may be more subtle in most places. Wintertime and nighttime temperatures will warm more than summer daytime temperatures. This is because higher CO_2 acts to decrease radiative heat loss from Earth's surface to space. Greenhouse gases not only warm the planet in general, but they also hold the heat in longer during the cold times. All seasons tend to warm, but cool times get a double whammy.

Land tends to warm more than water because evaporation may carry away heat from the water, but the land may dry out. Some models predict a general drying out of continental interiors for this reason. Growing seasons will get significantly longer. Growing seasons are already about a week longer than they were a few decades ago.

There will be more days of extreme heat, and fewer days of extreme cold. Projections of mortality from heat waves show an increase in heat-related deaths. It must be said however that there are also projections of mortality from cold, which decrease. No doubt the residents of tropical cities like New Delhi, already roasting in tropical urban heat islands, will not welcome a further few degrees of heat. A two-week heat wave in Europe in August 2003 is estimated to have killed 35,000 people. Canadian farmers on the other hand may find advantage in the new climate, with longer growing season and milder winters. The projections of economic impacts of climate change show winners as well as losers for small changes in climate, while for large climate changes almost everyone loses.

The warming due to CO_2 is offset by cooling from sulfate aerosols, but the amount of cooling from the aerosols is not the same everywhere. The CO_2 radiative effect is pretty much the same everywhere, proportional to local temperature, because CO_2 is a well-mixed gas in the atmosphere. Aerosols, on the other hand, last only a few weeks before they are removed from the atmosphere as acid rain. The cooling effect of the aerosols is therefore concentrated near the sources of their release, predominantly in the industrialized northern hemisphere.

Rainfall

Warm air holds more water vapor than cool air, so the global rate of rainfall is expected to increase with warming. An increase in rainfall overall sounds like a good thing, in a time when freshwater availability is a problem in many parts of the world. However, for the rainfall forecast the devil is in the distribution, even more than it is for temperature (Plate 12.3).

Part of the pattern of rainfall change has the appearance of the Hadley circulation (Fig. 12.5). Air rises convectively at the equator, where the solar heating is most intense. The rising air cools, condensing water into rain. Air subsides in the subtropics, bringing bone-dry air down to the surface and creating areas of minimal rainfall at about 30° latitude North and South. The anthropogenic impact is expected to increase the Hadley circulation pattern, intensifying rainfall at the equator and further drying out

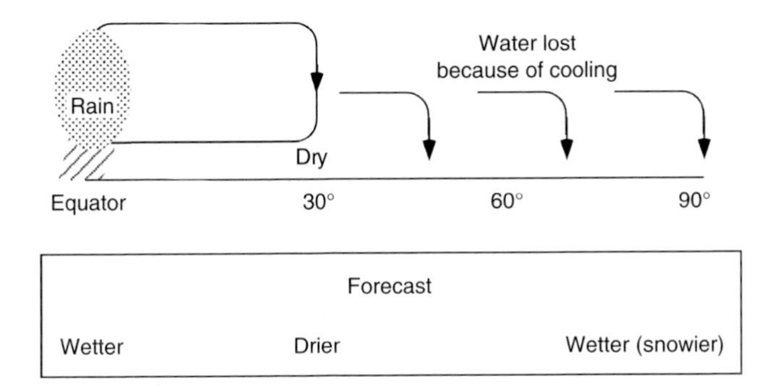

Fig. 12.5 **The Hadley circulation and the rainfall forecast.**

the desert regions in the subtropics. Meanwhile, precipitation in the highest latitudes is expected to increase because of the increase in temperature (remember there is more water vapor in warm air). This increase in high-latitude precipitation plays a role in the projections for the large ice sheets (next section).

Several facets of the rainfall forecast point toward an increased danger of drought. The dry subsidence areas of the Earth, the global desert belts as 30°N and S latitudes, are expected to dry further. Droughts of just a few years are sufficient to upend the lives of millions. With a global change to a new climate, there is the possibility for regional droughts that could last for centuries, like the fifth-century American Southwest drought during the Medieval Warm period in Europe.

Ice

For our purposes we can divide the frozen water on the planet into several types. ***Sea ice*** forms from seawater freezing at the surface and may be augmented by snow that lands on the ice. ***Glaciers*** are formed from compacted snow on a mountain-top, and flow down mountains like silent rivers of ice (Fig. 11.4). ***Ice sheets*** are larger than glaciers and are found today in Greenland and Antarctica. ***Ice shelves*** are former ice sheets that have flowed out onto water, floating but still attached to the ice sheet.

The fate of the major ice sheets depends on the balance between snowfall and melting. For the ***Greenland ice sheet***, the air around the base of the ice sheet is warm enough that a temperature increase leads to a significant increase in melting. Warming of about 3°C over Greenland over present-day temperatures is thought to be enough to ultimately melt the Greenland ice sheet. Recall that warming is not only more intense in high latitudes than the global average warming from increasing CO_2, but also more intense in winter than in summer. The ice sheet is mostly sensitive to summer temperature, when melting takes place. Models predict that the Greenland ice sheet will eventually melt if the temperature were 3°C warmer than today. The real question is how quickly can it happen (see Sea level section). The Greenland ice sheet could increase sea level by up to 7 m if it were to melt.

The ***Antarctic ice sheet*** is colder than the Greenland ice sheet, and an increase in temperature is not predicted to have as strong an impact as it will for Greenland. Snowfall is expected to increase in the Antarctic interior because more water vapor will be carried in the warmer atmosphere. Overall, the Antarctic ice sheet is expected to increase in size over the next century; this is consistent with present-day observations. A piece of the Antarctic ice sheet has the potential to melt down catastrophically, however. This is called the ***West Antarctic ice sheet***. The base of this ice sheet rests on rocks below sea level. Portions of the ice itself are flowing at speeds of hundreds of meters per year in rivers of flow called ***ice streams***. There is geological evidence to suspect that the West Antarctic ice sheet may have melted in the past, over time periods of centuries, for example during the transition from glacial to interglacial climate around 12,000 years ago. The general consensus is that the West Antarctic ice sheet is not likely to melt down significantly in the coming century, but if it were to melt, it could increase sea level by about 5 m.

The area of the ocean surface that is covered by sea ice is predicted to decrease. Models predict that the sea ice in the Arctic Ocean may melt or exist only seasonally within the coming century. Ice cover in the Arctic has decreased to the point that the mythical Northwest Passage may finally become a reality, at least in summer. This is an enormous change in the albedo, and therefore the energy balance, of the high northern latitudes. Polar bears are sure to become extinct if the ice melts (see Biological effects section).

Glaciers are expected to melt with warming, indeed have already begun doing so (see Fig. 11.4). This water contributes to sea level rise (next section). Perhaps the most pressing concern about the loss of glaciers is that they supply water through the summer to a lot of people (see Human impacts section).

Sea level

Sea level is expected to rise with global warming, for two reasons. One is the thermal expansion of water in the ocean. It takes a thousand years for water to cycle through the deep ocean, so this is how long it will take to warm the entirety of the ocean. Sea level from thermal expansion will also take this long. Depending on the amount of warming, and the circulation and mixing in the ocean, the IPCC forecast calls for 0.1–0.4 m of sea level rise due to thermal expansion alone by the year 2100. Models suggest that a thermal expansion from long-term CO_2 doubling might eventually, after a thousand years or so, raise sea level by 0.5–2.0 m.

The other component of sea level rise is from melting ice, but only ice that was not previously floating. Sea ice and ice shelves, when they float in water, displace their weight in water. This is called ***Archimedes' principle***. Archimedes is said to have discovered this property while floating in a bathtub, and was so excited that he ran down the streets of Athens naked yelling "Eureka!" (I found it!). When floating ice melts, its water exactly fills the hole that the floating ice had occupied. Land ice, glaciers, and ice sheets do increase sea level when they melt. The melting of land ice also affects the gravitational field in the area, and astonishingly, the gravitational field has an even stronger effect on local sea level than does the fact that the ocean has more water in it after the ice melts. The melting of an ice sheet actually depresses sea level in its immediate previous vicinity, but raises sea level everywhere else.

The forecast for the next century is for 0.1–0.7 m of sea level rise, approximately half due to expansion and half due to melting of glaciers and ice sheets. One meter of sea level rise inundates about 10^{11} m of land, about 0.07% of the land area of the Earth. The roughly half meter of projected sea level rise is small compared with the range of tidal variation in most places, so if sea level were raised by a half meter today, it's not clear that it would be an immediate disaster right away. The impacts would be focused at particular times, such as storm surges, and particular places, like Bangladesh, the Nile delta (Egypt), the Mississippi delta (New Orleans), New York, Miami, and small islands in the Pacific Ocean.

Sea level at any given location on the Earth depends not only on the total volume of water in the ocean, of course, but also on local geological movement of the ground, up or down. The largest vertical motions on the surface of the Earth today are the result of melting of ice sheets from glacial time. Remove all that weight and the crust will float higher relative to the surrounding crust (Eureka!). There is also vertical motion in other parts of the world which is comparable to the projected rate of sea level rise. Bangladesh is a sad example; the local sinking or subsidence rate of the land there is about 0.5 cm/year, comparable with the projected rate of sea level rise, doubling the inundation that Bangladesh would experience naturally.

It may be possible to defend some of the land against the invasion of the ocean. Much of Holland is located below sea level, and the ocean is kept at bay with dikes and pumps. This could presumably work for Manhattan and New Orleans as well. For Bangladesh and Egypt, where the coastlines are longer, it may be less tractable. A sea level increase of 1 m could inundate about 10% of the farmland in these countries, displacing millions of subsistence farmers. The first impacts of sea level rise are felt by the farmers. The Pacific island of Tuvalu imports food because they can no longer grow their own.

There are indications from recent events and from the geologic record to worry a bit that the consensus view about slow ice sheet melting might be overly optimistic. Ice shelves have recently demonstrated a dramatic ability to self-destruct catastrophically. A piece of the ***Larsen ice shelf*** on the Antarctic peninsula, the size of Rhode Island, was there during one pass of a satellite, and had exploded into what glaciologists call "blue slurpie" for the next pass of the satellite several days later (Plate 12.4). This explosion can be explained as the result of meltwater at the surface creating crevasses, canyons in the ice, filled with water, cutting the ice shelf into blocks like dominoes standing on end (Fig. 12.6). If the blocks are taller than they are wide, they will have a tendency to tip sideways, expelling the rest of the blocks out to sea and provoking them to tip as well. Kablooey.

During the transition from glacial to interglacial climate 14,000 years ago, sea level rose 5 cm/year, 10 times faster than our forecast, for 400 years. It is not clear exactly where this water came from, it could have been real-time melting or it could have been trapped meltwater in a glacial lake, dammed by ice. A bit earlier than that, during the depths of the glacial climate, sediments of the North Atlantic Ocean tell us about

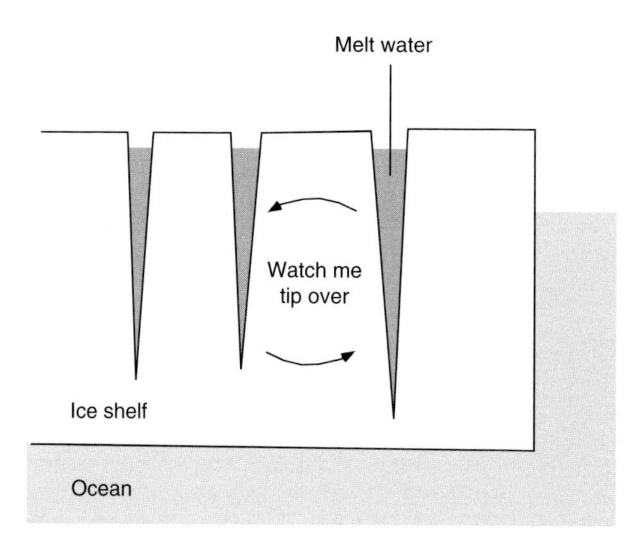

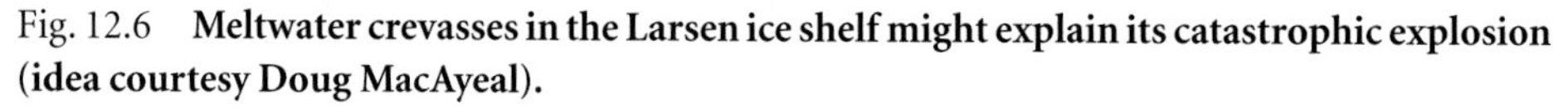

Fig. 12.6 **Meltwater crevasses in the Larsen ice shelf might explain its catastrophic explosion (idea courtesy Doug MacAyeal).**

catastrophic ice collapses during glacial time, called ***Heinrich events***. The sediments contain layers of rocks (***Heinrich layers***) that are too big to be carried by wind and could only have been carried in floating ice, enormous armadas of icebergs spewing out to melt in the North Atlantic for centuries. The rate of sea level rise associated with Heinrich events has been estimated to have been 1–10 cm/year for hundreds of years. The mechanism by which these icebergs were launched remains a mystery, but launched they were, as ice not as any dammed meltwater. Perhaps some mechanism such as the Larsen ice shelf collapse was responsible.

The significance of iceberg discharge is that it is an extremely efficient method of melting ice. Melting of an ice sheet in place is slow because it takes a long time for the atmosphere to carry that much heat up there. Icebergs move the ice mountain to the Sun.

The melting of an ice shelf does not contribute to sea level rise because the ice shelf was floating already. The concern is that the ice shelf might have acted to stopper the flow from the ice sheet on land, uphill from it. Take away the ice shelf and flow from the ice sheet might accelerate. It takes more ice than what is in an ice shelf to create a Heinrich layer, so ice sheets have demonstrated the ability to collapse into the ocean before, even if we don't understand very well how they do it. Several of the ice streams draining the West Antarctic ice sheet run into the Ross ice shelf, which is itself within a few degrees of melting like the Larsen ice shelf did. The West Antarctic ice sheet, recall from the Ice section, seems precariously perched on bedrock below the sea floor.

The long-term geologic record is alarming (Fig. 12.7). The temperature of the Earth was perhaps 4–5°C warmer than today 40 million years ago when the Antarctic ice sheet first made an appearance. Sea level was about 70 m higher then than today. During glacial climate, temperature was about 6°C colder than today, and sea level 120 m lower. If we connect the dots, we see that a huge eventual change in sea level

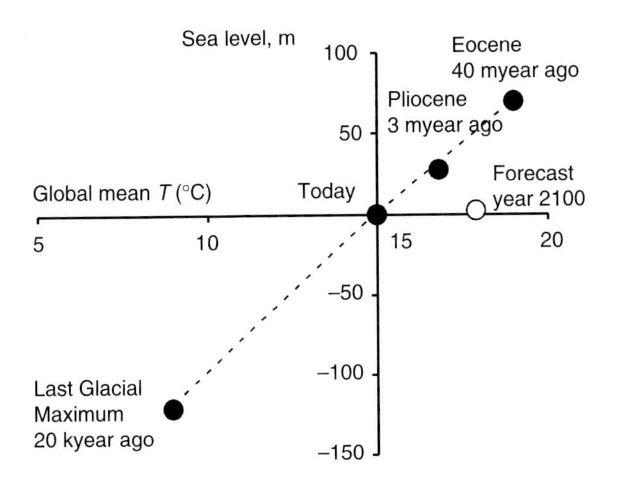

Fig. 12.7 **The relationship between sea level and temperature on geologic timescales. The IPCC forecast is for 0.5 m, whereas the geological forecast is for 50 m, eventually.**

might accompany a long-term increase in Earth's temperature. The question is how quickly the ice melts.

Floods

The climate forecast warns of an increased risk of floods. One reason is that global rainfall is predicted to increase, as we saw at the Rainfall section and in Fig. 12.4. Another reason is that there is a correlation in our weather between the average amount of rainfall at a place, and the variability in rainfall at that place. More rain begets more hard rain. For this reason, the frequency of extreme rain events is expected to increase with warming. Extreme rain events lead to floods.

The variability from one year to the next of the monsoon in India is expected to increase. This may lead to floods in high monsoon years, and droughts when the monsoon is weak.

Hurricanes also generate floods; Katrina in New Orleans being a recent infamous example. It rains a lot during a hurricane. Hurricanes also elevate the sea surface into what looks like a very high tide called a ***storm surge***. These are caused by the low atmospheric pressure inside a hurricane lifting up the sea surface. Very strong Category 5 hurricanes are expected more frequently (next section), and sea level is rising (last section), a double whammy for flooding. This trend is apparent in data from the real climate, and seems like a strong prediction for the future as well.

Hurricanes

Rising sea surface temperatures could conceivably fuel an increase in the number or the intensity of tropical cyclones (also called typhoons or hurricanes), because warm

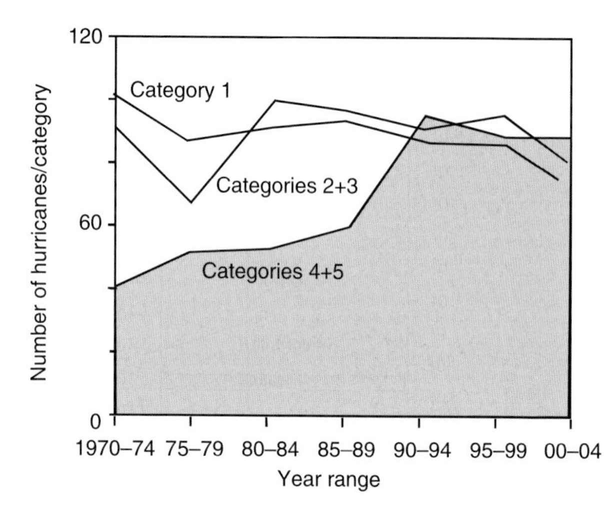

Fig. 12.8 **The frequency of different intensities of hurricanes, from Webster et al. (2005).**

surface waters are the source of energy for these storms. Hurricanes do not form if the sea surface temperature is cooler than 26°C. As the sea surface temperature increases, the maximum possible intensity goes up. Cyclone intensity also depends on the temperature at the top of the atmosphere, however, which goes up more or less in parallel with sea surface temperature, counteracting some of the effect of the sea surface temperature rise. Cyclone intensity depends on the way that winds are blowing with altitude, whether they tear a tropical storm apart or allow it to grow into a full cyclone. The evolution of a cyclone depends on the subsurface temperatures of the ocean, also, because the wind tends to mix warm surface waters with the cooler waters of the subsurface, slowing things down. The bottom line is, there are reasons to worry that cyclone might get more intense with rising CO_2, but the connection is not as simple as, say, the link between CO_2 and global mean temperature.

Every year there are about 90 tropical storms, 40 of which tend to become cyclones. A cyclone requires a jump-start, some chance combination of winds and pressures that enables the tropical storm to get going. One might guess that the number of tropical storms would depend on climatic conditions; el Niño or warming or something like that. However, the frequency of tropical storms is observed to be fairly constant through time (Fig. 12.8). No one knows why.

The intensity of hurricanes, on the other hand, seems to be climate sensitive. Model simulated hurricanes are more intense in a doubled-CO_2 world. Ordinary climate models do not have enough resolution, meaning that their grid points are not close enough together, to simulate hurricanes very well. Hurricanes can be simulated in global models, however, by a technique called ***adaptive grid refinement***, essentially inserting grid points into a model when and where they are needed, or you can think of it as running a second, high resolution model in the area where a hurricane is, coupled to and exchanging information with a low-resolution global model. Using a wide range of models and for a wide range in parameter space, the study of Knutson and Tuleya (2004) found that the cyclones in their model shift to higher intensity in

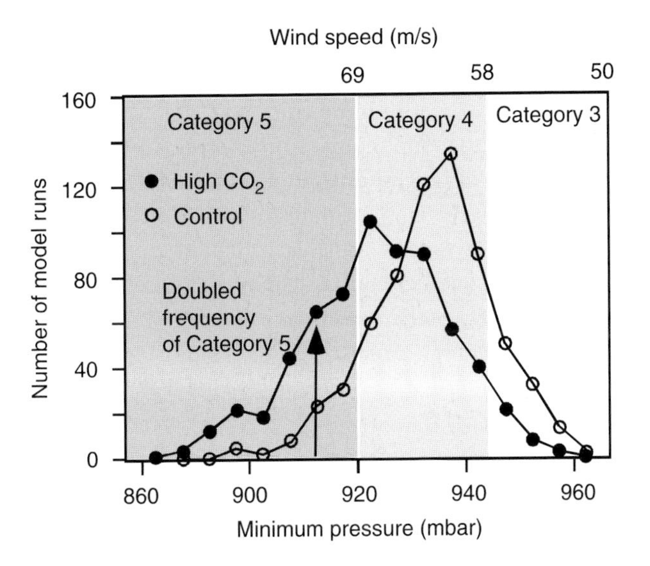

Fig. 12.9 **The distribution of wind speeds in model hurricanes for a control world (natural climate) and doubled CO_2.**

a high-CO_2 world (Fig. 12.9). Wind speeds increase by 10% or so. The destructive power of a hurricane increases, not linearly with the wind speed, but as the wind speed raised to the third power. So an increase in wind speed of 10% leads to an increase in damage of about 33%. The cyclones to be feared the most are not the average ones, but the highest intensity ones. The frequency of the highest power, Category 5, cyclones is predicted by the model to double or worse in a high-CO_2 world.

The frequency of the most powerful hurricanes, called Category 5 storms, has doubled in the past 30 years (Fig. 12.8). This by itself does not look to me like iron-clad proof that a global warming future would have more intense cyclones because there is natural variability in these things. A more compelling piece of evidence is that the total destructive power of cyclones in the past 50 years has correlated extremely tightly with variations in sea surface temperature (Fig. 12.10). The cyclones are getting stronger even than the models or theory say they should, given the amount of warming that the sea surface has undergone. It is dangerous to extrapolate this trend to the future until we understand the observation better. However, the 2–3°C of sea surface temperature warming we might expect in the future is much larger than the 0.5°C of warming we've had already, and cyclones have doubled their power already. This gives reason for pause, even if we don't know for sure what's going on.

The really provocative observation is a massive increase in the cost of weather-related damages in the past 50 years (Fig. 12.11). The number of weather-related events has increased by a factor of five since the 1950s. The number of nonweather related disasters has gone up also, but only about half as quickly as weather-related claims. Insurance payments for weather-related property destruction have increased by a factor of about 13 in the past 50 years. A trend of rising affluence and coastal construction can

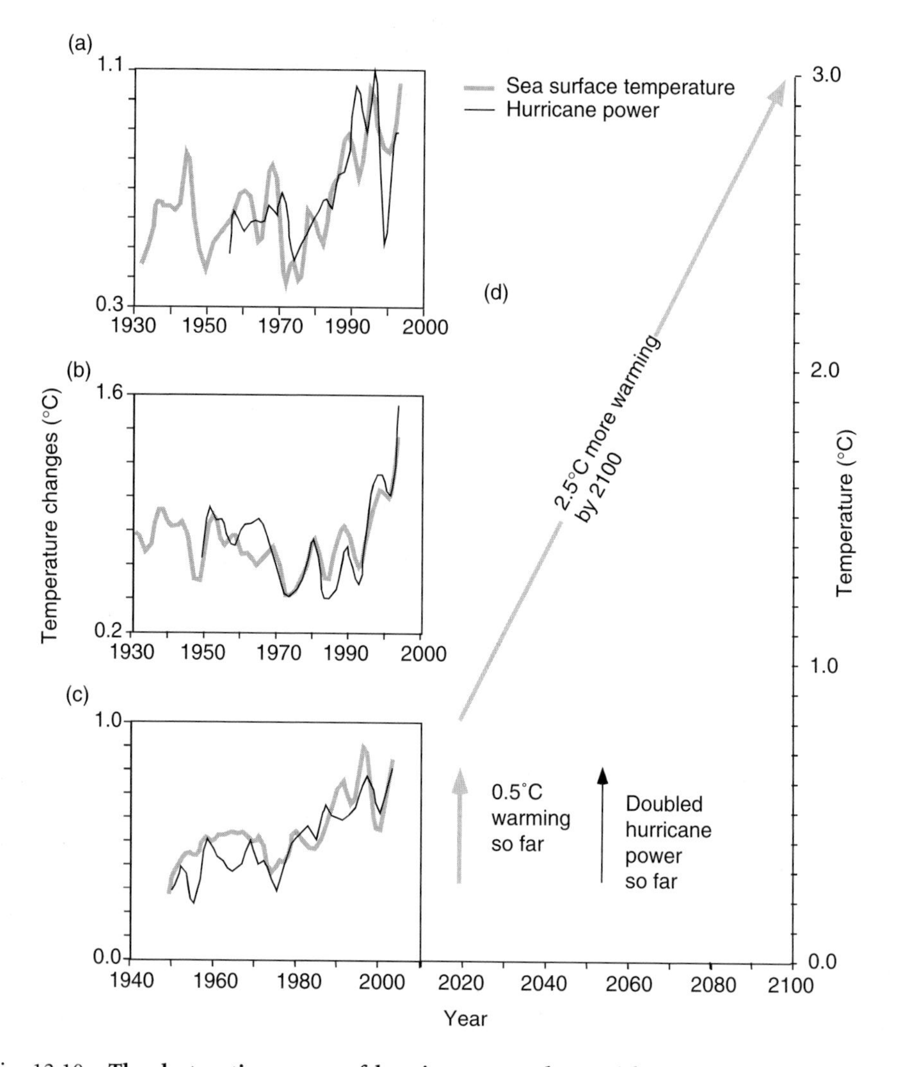

Fig. 12.10 **The destructive power of hurricanes correlates with variations in sea surface temperature, in (a) the North Pacific, (b) the North Atlantic, and (c) both basins combined, from Emanuel, (2005), (d) is a possible future sea surface temperature rise. *Disclaimer*: There is no hurricane power prediction on the right-hand part of the plot because we do not know what the hurricane response will be. The simplest assumption would be that the response in the future will be proportional to the response in the past, which would be huge given the huge future temperature changes. But it might not work like that. The point is that the future temperature changes could be huge compared to what we have seen so far.**

explain at least part of the increase in payouts, perhaps all of it. Some of the damage is due to floods, which are probably caused by land-use changes, cutting forests, draining wetlands, and building in flood plains, as much as by climate variability. A number of sources, including Working Group II of the IPCC (the human impacts group), blame the increase in insurance payouts on climate change, but the observational support for that conclusion is thin.

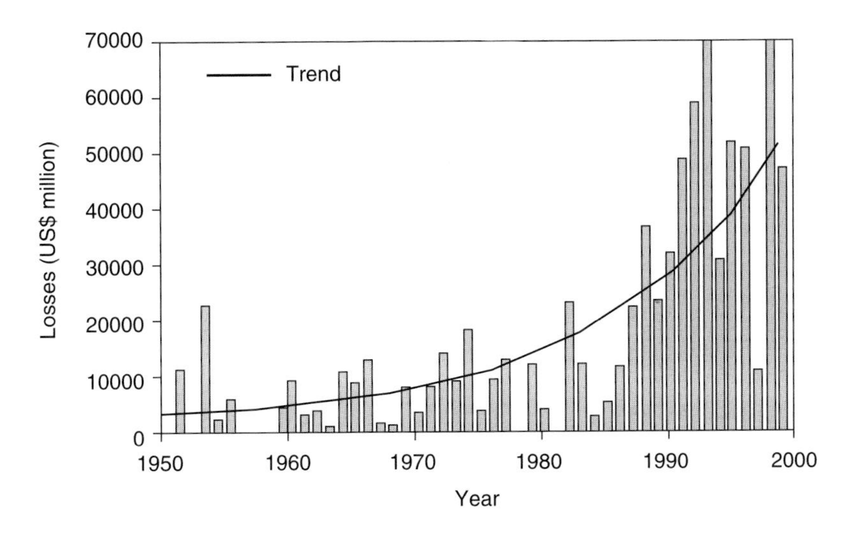

Fig. 12.11 **Economic losses (insured + uninsured) due to weather-related disasters, from IPCC (2001).**

Biological impacts

The character of the land surface, whether it is hardwood forest or grassland for example, is determined to a large extent by climatic forcing. Temperature is important of course, although it may be the coldest temperature of the year, or the warmest, rather than the annual average. Precipitation also plays a huge role in shaping the natural landscape.

The infinite complexity of the real landscape can be modeled by simplifying it into some manageable number of different biome types, such as "tropical deciduous" or "savanna" (Plate 12.5). Biomes can be modeled by keeping track of the essentials, the budgets of energy, water, and nutrients for example. Different biomes have different capabilities and tolerances, and they compete with each other for land surface area.

The equilibrium biome response to a new climate can probably be predicted fairly confidently because the distribution of biomes on Earth today is fairly well correlated with climate. The simplest model for terrestrial biomes could simply be a look-up table of the ranges of conditions inhabited by each biome type.

It is more difficult to forecast transitions between one climate state and another. When climate changes, the optimal biome type for some spot on Earth might change. In mountainous areas, biomes may be able to move to higher elevation where it is cooler (remember the lapse rate?). In flat places, the climate shift could be hundreds of kilometers. Some organisms, like insects, move immediately, but for others, like trees, the story is a bit more complicated. Of course we are not talking about walking trees, we are talking about seeds and new trees, and a process called ***ecological succession***. When a forest grows after a clear cut, for example, the first set of plant and tree species will be different from the species that you would find in an old-growth forest under the same climatic conditions. The process of ecological succession is not trivial to understand in the real world, which makes it even more difficult to forecast for the future.

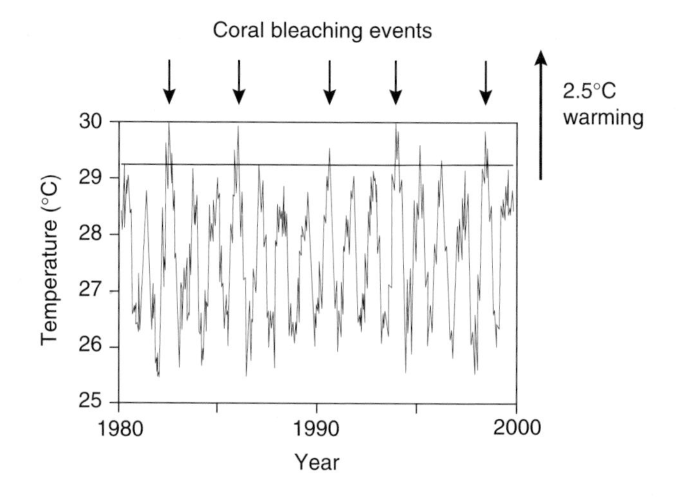

Fig. 12.12 **Coral bleaching events in Tahiti. Bleaching events occur whenever temperature exceeds a threshold value. A future warming of 2.5°C would be very large compared with the coral stress level. From Hoegh-Guldberg (2005).**

The biotic response of the climate model from before is shown in Plate 12.5. The most striking change in the Earth's landscape is in the Arctic. The biome response is strong there because the climate change is intense because of (let's all say it together) the ice albedo feedback. The melting permafrost described above also changes the biotic landscape. The model in Plate 12.5 experienced a near total loss of the tundra biome by the year 2300, and the loss of most of it by 2100.

Melting of sea ice in the Arctic has the native polar bear on the brink of extinction. Polar bears only eat in the winter, by fishing for seals through holes in the ice. In summer, when the ice near the coast melts, the polar bears have to live on their accumulated fat. With further melting of Arctic ice, the extinction of natural polar bears is virtually certain.

In the ocean, ***coral reefs*** appear to be particularly vulnerable. Reefs are built by sedentary animals, the corals, that house symbiotic algae, plants that can perform photosynthesis to aid the nutrition of the corals. Coral reef ecosystems are among the most diverse on Earth. Reefs of one type or another have been found throughout the fossil record, yet coral reef systems today appear to be rather fragile. If the water is not clear enough, the symbiotic algae will not get enough light, and the coral will suffer. Corals build the reef from $CaCO_3$, the production of which becomes more of a challenge as the seawater becomes acidified by rising CO_2 concentrations. Overfishing threatens the coral community, in particular fishing by means such as dynamite or cyanide poison as practiced in some parts of the world. Corals have also been ravaged by diseases and by invasive species.

In addition to all of these struggles, corals are vulnerable to increases in temperature (Fig. 12.12). When corals are stressed, they respond by expelling their symbiotic algae. This is called coral ***bleaching*** because of their loss of color. Bleaching may be a mechanism for the coral to try to find new symbiotic algae that are more suited to the conditions they find themselves in, but it is a measure of desperation. Bleaching is often followed by the death of the coral. Figure 12.12 shows a record of temperatures in Tahiti, with arrows indicating times of coral bleaching events. The correlation between heat spikes and bleaching is very clear. Sediment cores in Caribbean reefs indicate that the degradation of reef communities is more widespread than has been seen in centuries. The projected warming in the future looks like it will have catastrophic effect on corals.

In general, the diversity of life (number of species) on Earth is already decreasing because the natural world is restricted and fragmented by human land use. Climate change can only amplify this extinction trend by demanding that natural ecosystems get up and move, just when they are hobbled and unable to do so.

Carbon cycle impacts

CO_2

We learned in Chapter 10 that the carbon buffer chemistry in seawater loses its capacity to uptake CO_2 as the CO_2 concentration rises. As the temperature rises, changes in the ocean circulation may decrease the rate at which CO_2 is carried into the subsurface ocean, slowing the CO_2 drawdown from the atmosphere.

One thing we would very much like to predict using a biome model is whether the terrestrial biosphere will take up CO_2 in the future or release it. Today, the land surface overall appears to be a wash. There is CO_2 released from tropical reforestation, while an equal amount appears to be taken up in high-latitude northern forests. If temperatures increase too much, though, it is possible that faster respiration rates in soils may tip the balance toward net releasing CO_2. Respiration converts organic carbon back to CO_2, and in soils this is done by bacteria and fungi. There is a clear trend between temperature and soil organic carbon in soils around the world today.

The terrestrial biosphere model in the Projects section took up about 500 Gton out of a total fossil-fuel CO_2 release of 5000 Gton C. There was also an increase in the soil carbon pool of about 1250 Gton C. Other models predict a net release of carbon (Cox et al. 2000). The role that the terrestrial biosphere will play is one of the major uncertainties in the climate change forecast.

Methane

As we learned in Chapter 9, there is an immense amount of methane frozen in microscopic icy cages called clathrates. There may be 100 times more methane in permafrost soils as there is in the atmosphere, and the ocean contains perhaps 1000 times as much methane as the atmosphere. If these deposits were to melt all at once, releasing the methane to the atmosphere, it would very bad, especially for the first decades, the atmospheric lifetime of methane. Increasing the atmospheric methane concentration by a factor of 10 would be about equivalent to doubling CO_2. (I learned this from the IR light model that we ran in Chapter 4.)

The good news is that no one has thought of any scenario for releasing a sizable fraction of all of this methane in less than a decade. Melting permafrost in Siberia is releasing methane to the Arctic Ocean, where it escapes to the atmosphere, but the melting of the permafrost will take decades or centuries. There is evidence of submarine landslides and sea floor explosions leaving what are called pockmarks, but the largest of these would have released perhaps 1 Gton C of methane, which would have a small climate effect by itself.

We learned in Chapter 10 that ongoing, chronic release of methane is what determines the ongoing methane concentration of the atmosphere. If we doubled the methane source flux, we would double (or perhaps a bit more than double) the atmospheric concentration. The sources of methane to the atmosphere have already increased enough since 1750 to double the atmospheric methane. Melting permafrost will release methane from clathrates, and thawing organic matter like peat and wooly mammoths will decompose to release more methane.

The increase in atmospheric methane has a significant effect on the climate of the Earth. Reducing methane emissions may be an economical way of avoiding some climate change. However, none of the models of the methane cycle, or melting clathrates, predicts that methane will have as much impact on climate as anthropogenic CO_2 will. When methane degrades, it converts to CO_2 and continues to warm the planet. Methane, molecule per molecule, is more powerful as a greenhouse gas (Chapter 4) but it is a transient molecule whereas CO_2 accumulates (Chapter 10). While the methane is being released, models often find that you get more radiative forcing from the accumulated CO_2 from past methane emissions than you do from the amount of methane in the atmosphere at any given time.

A model constructed by your author predicts that ultimately, after thousands of years, the methane reservoir could release as much carbon as we do by fossil fuel combustion (Archer & Buffett 2005).

Human impacts

Water availability may be the most important impact of global warming on human welfare. We have already discussed the possibility for regional drought in the global warming future. Climate rearrangement may generate regional droughts that could last centuries, as occurred in the American Southwest during medieval times. In addition, people in many parts of the world depend on melting glaciers and mountain snow pack as a source of summertime freshwater. Such places include along the Himalayas in India, Pakistan, and Uzbekistan, along the Andes in Lima, La Pas, and Quito, and along the Sierras in the North American Pacific Northwest.

The impacts of climate change to agriculture are potentially large. There are projected changes in crop yield, divided by crop and by region, of order 10–50%, in response to future climate change predictions for the coming century (IPCC 2001 Working Group II report, Impacts, Adaptation, and Vulnerability http://www.grida.no/climate/ipcc_tar/).

In general, negative impacts are stronger in the tropics and in the developing world, whereas the higher latitude temperature countries suffer less or may even benefit.

There are several factors at work here. One is the increase in temperature. Corn yields are sensitive to heat spikes; just a few can impact the bottom line for the entire season. Water availability is also a primary controller of crop yield. Dry regions are expected to be even drier with global warming, decreasing crop yields in these areas. On the other hand, increasing CO_2 concentration in the air is expected to boost agricultural production by the CO_2 fertilization effect (Chapter 10). Higher CO_2 levels help plants deal with water stress by allowing them to get the CO_2 they need for photosynthesis without opening their stomata as much. Another factor that is important in the forecast is adaptation. Perhaps if corn no longer grows in a region because of more frequent heat waves, the farmer will plant wheat, or some genetically modified Frankencorn that is more heat tolerant.

In general, the story of agriculture in the past century has been explosive, with production growing even faster than population, at ever-diminishing costs. This agricultural revolution is in large part due to industrial production fertilizers like nitrate and ammonia from atmospheric nitrogen, an energy intensive process. The explosive growth of agriculture in the recent past tends to make any worries about a collapse in global food production seem a little silly.

Regionally, the picture looks a little scarier. Perhaps in an age of globalization, it makes less difference where on Earth our food is grown because it can be transported around to where the people are. Or perhaps a rearrangement of rainfall and agricultural productivity could lead to famine, refugees, and war.

There is a concern that rising temperatures will widen the area where tropical diseases such as ***malaria*** and ***dengue fever*** will spread. Both have a significant impact on public health, and both appear to be spreading with an increase in the tropical zone of the Earth. It must be said here that malaria could be fought much more effectively than is currently being done everywhere, with mosquito nets and health care. I guess my personal feeling is that the spread of tropical diseases is a serious threat, but it is not the first and most important reason to combat global warming.

Abrupt climate change

In one way, the IPCC forecasts represent the best-case scenario because they are smooth. There are no surprises. The climate record from Greenland shows a period of stability throughout the Holocene, the past 10,000 years when civilization and agriculture developed (Fig. 12.13). Prior to this, during the last ice age 20–80 kyear ago, the climate of Greenland was not so stable. Temperatures varied wildly in 1000 year swings called ***Dansgaard–Oeschger events***, sudden warmings of 10–15°C within a few years, followed in general by more gradual coolings check. The record is punctuated by what are known as ***Heinrich events***, times when an ice sheet collapses into the ocean within a few centuries. These climate events are most intense in the high Northern latitudes, but their effects are felt as far away as New Zealand and Antarctica.

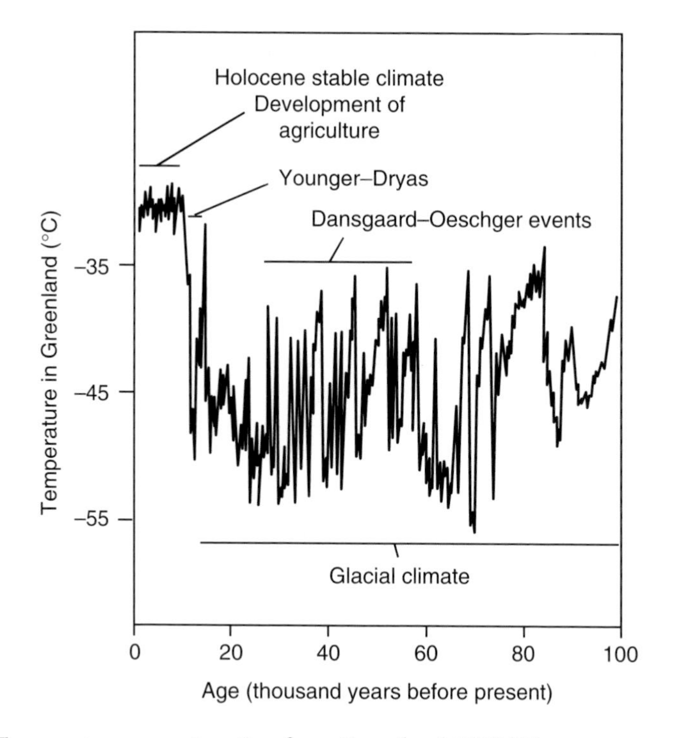

Fig. 12.13 **Temperature reconstruction from Greenland GISP II ice core.**

Scientists are still trying to figure out the reasons for all of the variability in the glacial climate. The circulation in the North Atlantic is one player. The Atlantic carries a considerable amount of heat into the high latitude Northern hemisphere. The warm surface waters in high latitude cool and sink, making way for more warm surface flow, bringing in more heat.

During glacial time, climate seemed able to rearrange itself into a different configuration. In this configuration, ice covered much of the North Atlantic. The overturning circulation in the North Atlantic slowed. The ocean carried less heat into the high latitude northern hemisphere. These events were apparently triggered by excess freshwater release into the North Atlantic. There are examples of climate changes from spilling of a dammed lake and by Heinrich event ice sheet meltdowns.

Here's a scenario. The freshwater release created a layer of low-salinity water at the surface of the ocean. Water is less dense when it has less salt dissolved in it, so the lower salinity surface water could cool down all the way to freezing without becoming dense enough to sink. Now sea ice becomes a player in the drama. The albedo of the ice acts to cool the climate still further, at least in the northern high latitudes. The climate of Greenland is particularly sensitive to this climate rearrangement because it loses its maritime influence. Air temperatures over land or ice plummet to values much lower than would be possible over water.

The abrupt climate change scenario for the future that people have been fretting about may be a cousin to this cold configuration of glacial climate. There are two possible sources of excess freshwater to the North Atlantic. One is precipitation, which is expected to increase, because of rising temperatures and water vapor saturation and all that. The other source people think about is meltwater from Greenland. Perhaps the Greenland ice sheet collapses like a modern-day Heinrich event, releasing its freshwater into the North Atlantic over a century or so. In models, the overturning circulation in the North Atlantic can be stopped or greatly slowed by an excess input of freshwater. In coupled atmosphere ocean climate models, the overturning circulation in the North Atlantic slows down as you increase atmospheric CO_2. If we were to simply draw a parallel between the past and the future, the result could be a paradoxical cooling in the northern high latitudes. Climate models do not predict a huge climate impact from a North Atlantic overturning shutdown. However, climate models have a tendency to underpredict the swiftness and the severity of past climate variability that we know is possible in the real world. This may be because positive feedbacks in the real world are difficult to capture in models.

An abrupt climate shift would be a surprise, like the ozone hole. Once the climate rearranged into some different mode, it could conceivably remain locked in that mode for centuries.

Take-home points

1. Climate models that simulate the past predict huge climate changes in the future, relative to what civilized humanity has ever seen before.
2. It takes hundreds of years for the climate to respond fully to changes in greenhouse gas concentrations. Feedbacks amplify the climate impact, and they also prolong the timescale of the response.
3. There could be prolonged droughts, more powerful hurricanes, floods, and extinctions in store.

Projects

Point your web browser at http://understandingtheforecast.org/Projects/bala.html. The physics part of the model, temperatures and winds and currents and all that, was described in the Projects section of the last chapter. In addition to that, the run had a land biosphere model, developed at the University of Wisconsin in Madison. Plants on land are described by 16 different biome types, which compete with each other for the land surface. The biomes are listed in Plate 12.5. The run also used a simple groundwater scheme, much simpler than real soils. Instead of allowing groundwater to flow as happens in the real world, model soils collect water and allow it evaporate, like a bucket. If too much water accumulates, it will overflow into rivers and to the oceans.

You can see annual mean surface temperatures and mean surface temperatures for December through February, and June through August. You can also see precipitation, soil moisture content, and vegetation type.

Use the different data processing and visualization tools described in the Projects section in Chapter 11 to investigate the following questions:

a. What is the global average temperature increase from the beginning to the end of the simulation? Most climate simulations end at the year 2100, but this one goes to 2300. You may investigate either year as an "end of simulation."
b. What is the predicted warming in the winter high latitude, the summer high latitude, and the tropics?
c. Can you see evidence for poleward migration of biome types? How far do the biomes move?
d. Can you see a systematic change in precipitation or soil moisture with climate change?

Further reading

Emanuel, K., See *Divine Wind* (2005) about hurricanes. This is an elegant treatment of the history and physics of hurricanes written by the guy who, well, wrote the book on hurricanes. The book is replete with color photos, plots, and paintings, with a few equations and many quotes from antiquity.

For a street-level description of what climate change might be like, see *Mark Lynas' High Tide, How Climate Change is Engulfing our Planet* (2004).

Archer, D.E. and B. Buffett, Time-dependent response of the global ocean clathrate reservoir to climatic and anthropogenic forcing, *Geochem., Geophys., Geosys.*, *6*(3), (2005) doi: 10.1029/2004GC000854.

Bala, G., K. Caldeira, A. Mirin, M. Wickett, and C. Delira, Multicentury changes to the global climate and carbon cycle: results from a coupled climate and carbon cycle model, *J. Clim.* (2005), *18*, 4531–44.

Cox, P.M., R.A. Betts, C.D. Jones, S.A. Spall, and I.J. Totterdell, Acceleration of global warming due to carbon-cycle feedbacks in a coupled climate model, *Nature* (2000), *408*, 184–7.

Emanuel, K., Increasing destructiveness of tropical cyclones over the past 30 years, *Nature* (2005), *436*(7051), 686–8.

Hoegh-Guldberg, O., Low coral cover in a high-CO_2 world, *J. Geophys. Res. Oceans* (2005), *110*(C9).

IPCC Scientific Assessment, 2001, Cambridge University Press or downloadable from http://www.grida.no/climate/ipcc_tar/. Technical summary.

Knutson, T.R. and R.E. Tuleya, Impact of CO_2-induced warming on simulated hurricane intensity and precipitation: sensitivity to the choice of climate model and convective parameterization, *J. Clim.* (2004), *17*(18), 3477–95.

Kolbert, E., *Field Notes from a Catastrophe*, p. 192, Bloomsbury, 2006.

Stainforth, D.A., T. Aina, C. Christensen, M. Collins, F. Faull, D.J. Frame, J.A. Kettleborough, S. Knight, A. Martin, J.M. Murphy, C. Piani, D. Sexton, L.A. Smith, R.A. Spicer, A.J. Thorpe, and M.R. Allen, Uncertainty in predictions of the climate response to rising levels of greenhouse gases, *Nature* (2005), *433*, 403–6.

Webster, P.J., G.J. Holland, J.A. Curry, and H.R. Chang, Changes in tropical cyclone number, duration, and intensity in a warming environment, *Science* (2005), *309*(5742), 1844–6.

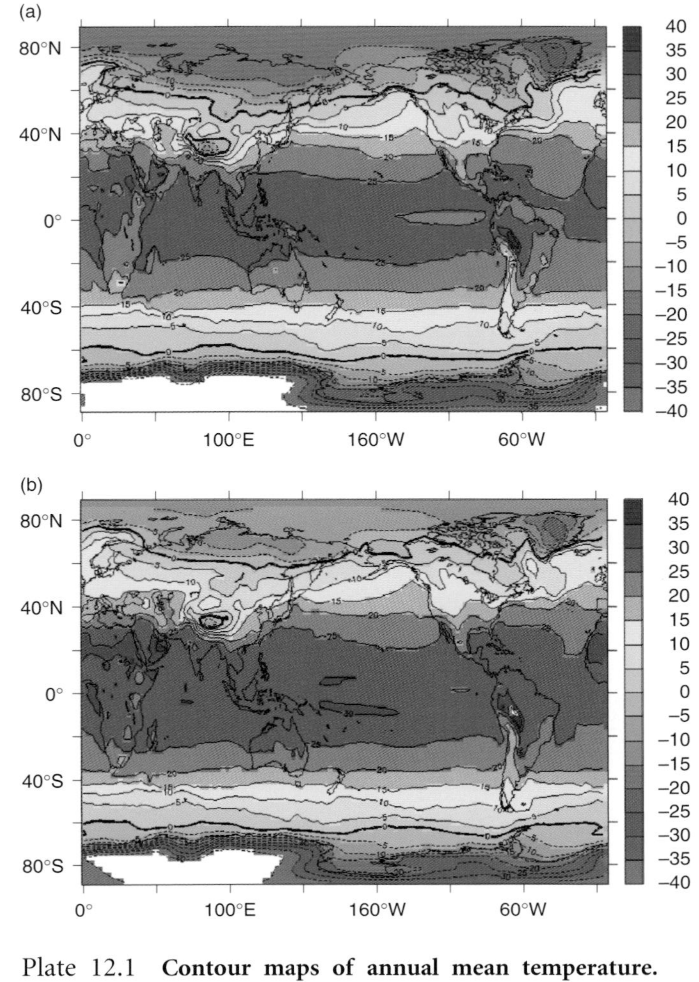

Plate 12.1 **Contour maps of annual mean temperature. From Bala et al. (2005). Temperature in (a) year 2000 and (b) year 2100.**

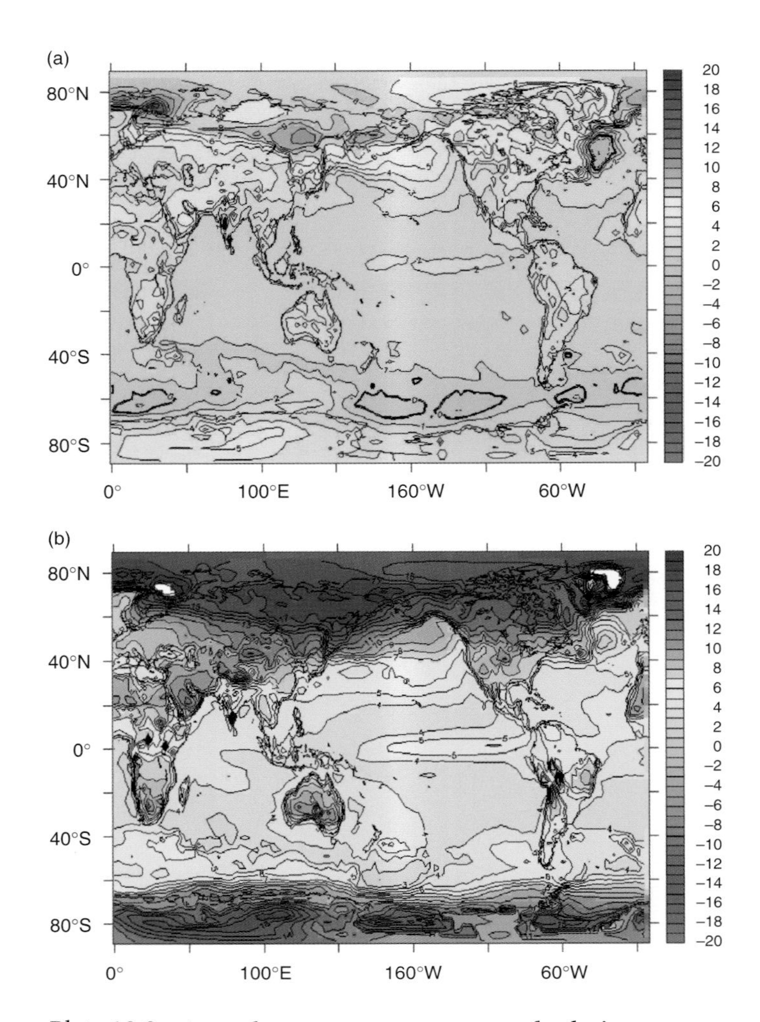

Plate 12.2 **Annual mean temperature anomaly, the increase from the year 2000. From Bala et al. (2005). Temperature rise °C from (a) 2000 to 2100 and (b) 2000 to 2300.**

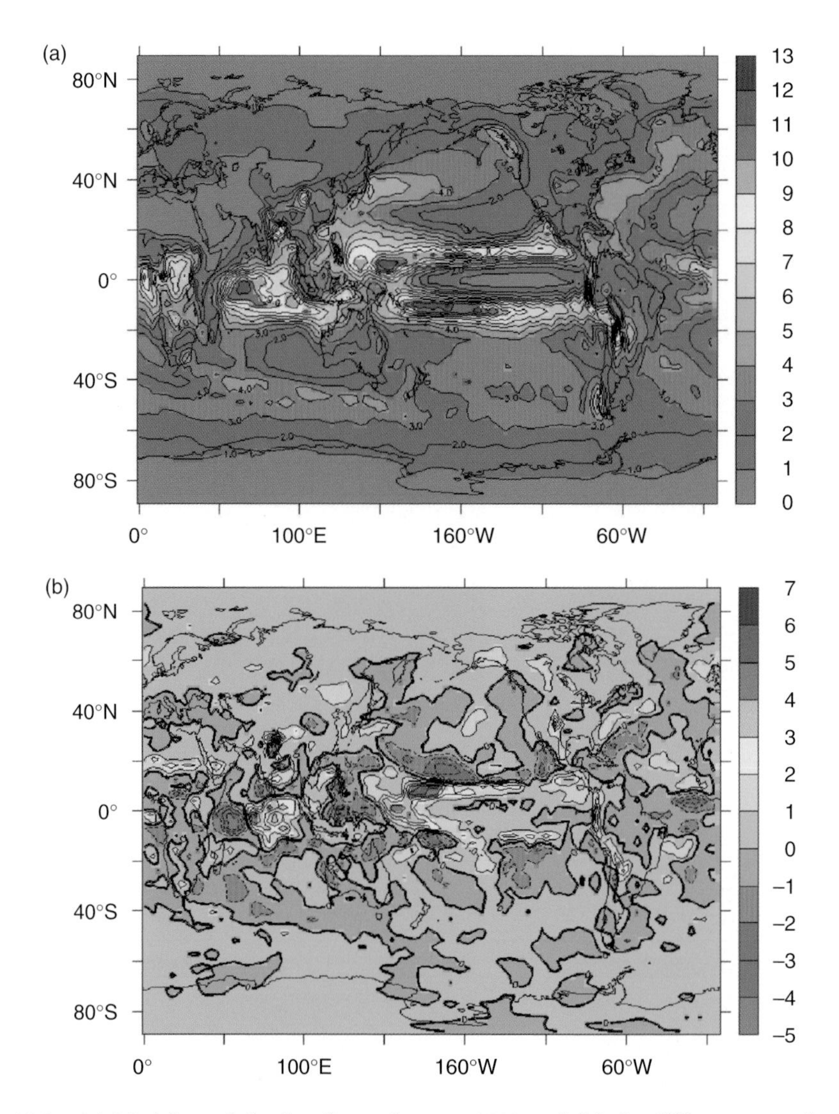

Plate 12.3 **(a) Model precipitation from the year 2000 and (b) the differences to the year 2100, from Bala et al. (2005).**

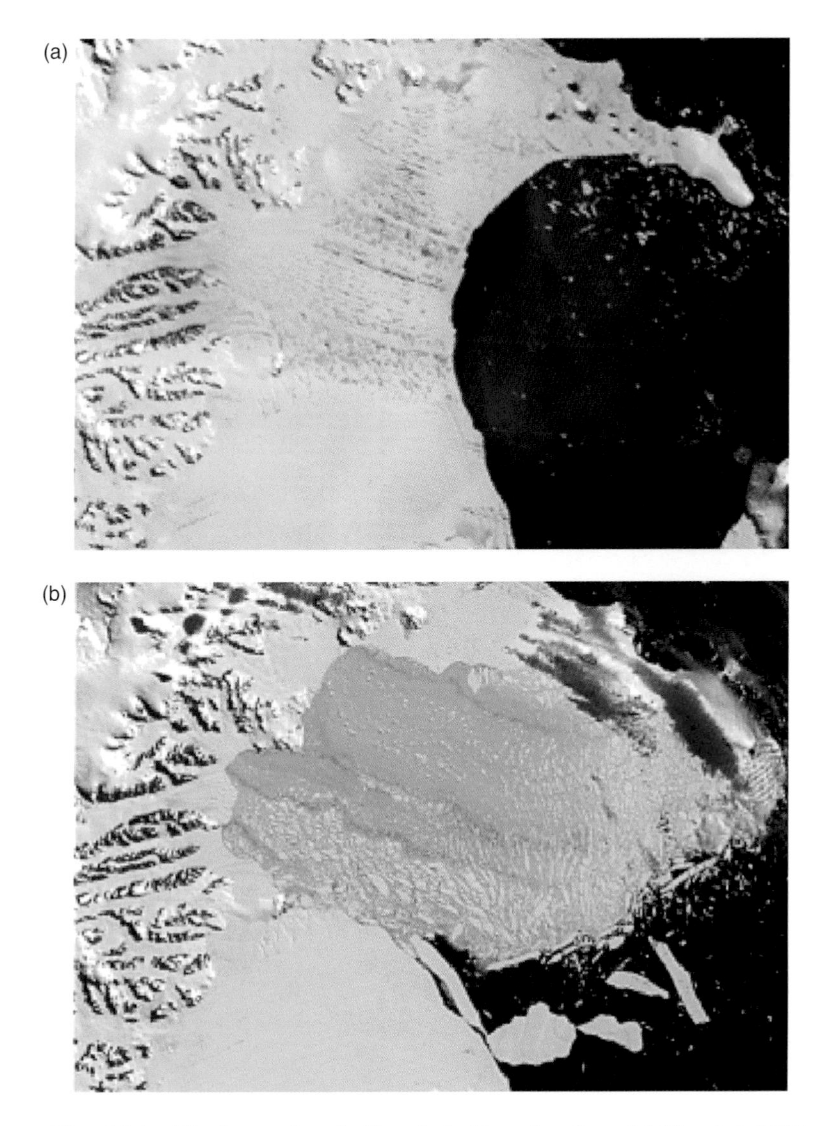

Plate 12.4 **The Larsen ice shelf exploded in a few days: (a) March 5, 2002 and (b) March 7, 2002.**

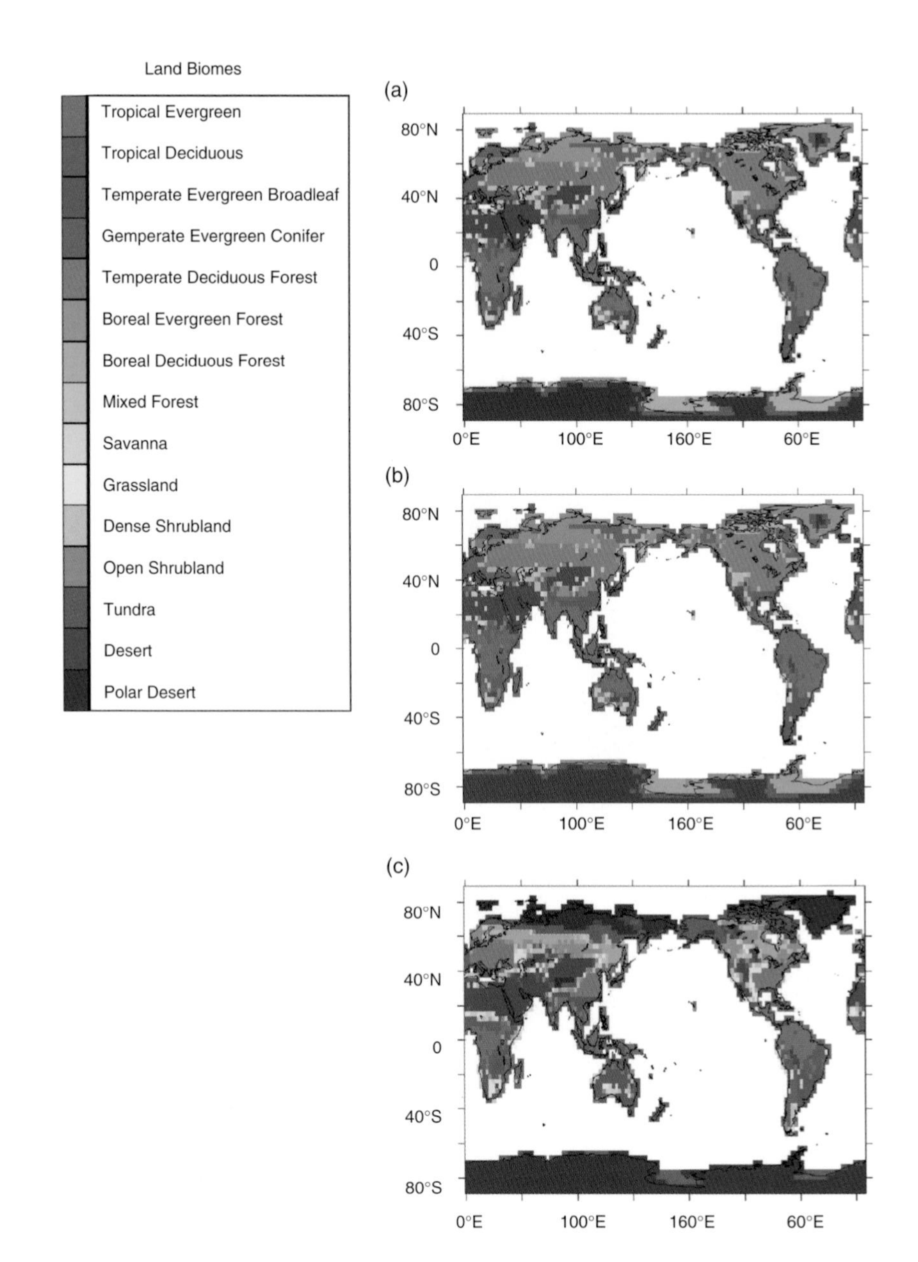

Plate 12.5 **Changes in terrestrial biomes, from the Bala et al. (2005) model. (a) Year 2000, (b) 2100, and (c) 2300.**

13
Decisions, decisions

Climate change is the largest-scale problem that mankind has ever faced, global in scope and extending across unimaginable stretches of time.

International negotiations began with the establishment of a framework within which to discuss, called the Framework Convention on Climate Change (FCCC). The FCCC is advised by the scientific body Intergovernmental Panel on Climate Change (IPCC). FCCC established an agreement called the Kyoto Protocol, an agreement to lower CO_2 emissions. The Kyoto Protocol is a good first step, but it can only be the first step in truly preventing climate change.

Economically, the annual costs of CO_2 emission limitation as specified by Kyoto are projected to be comparable with the costs of climate change. Since more than Kyoto will be required to truly prevent climate change, economics would seem to argue that it would be better to just live with climate change. However, economics is an awkward tool for making decisions about human lives and the natural world, especially decisions the consequences of which will reach far into the future.

Another way to approach the problem is to determine a danger level based on temperature or sea level, and ask how that danger level could be avoided. Perhaps we consider a sea level rise of more than 1 m to be dangerous, or a global mean temperature increase of 2°C.

No single energy source or strategy seems sufficient to solve the problem by itself, but by using a combination of new energy sources and strategies, it may be possible to avoid dangerous climate change.

Global warming is a large-scale problem

Global warming is a difficult problem to cope with politically because its footprint is so large. It is a global problem, subject to an effect known as the ***Tragedy of the Commons.*** An example of this effect is of a field for grazing sheep used in common by all residents of a village. Let's say that each new sheep added to the flock decreases the harvestable weight of each of the other cute little sheep by 10% (Fig. 13.1). A farmer with one sheep can add a second, doubling his number of sheep. But the result is that all sheep, including the farmer's, lose 10% of their harvestable weight (denoted as "lambchops" in Fig. 13.1). Each will provide 0.9 times as many lambchops as the original sheep did, so the farmer will now own the equivalent of 1.8 of the original sized sheep. Our farmer is better off than when he had only one sheep, while all the rest of the farmers lose 10% of their prior sheep stock.

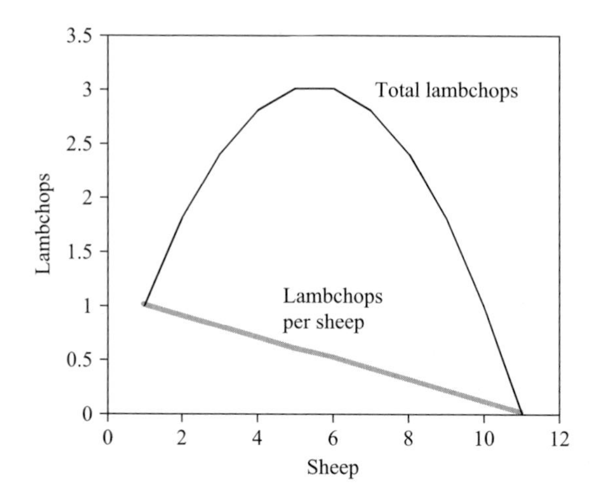

Fig. 13.1 **The tragedy of the commons demonstrated in a totally made-up story about sheep.**

In this particular example, there is an overall optimum number of sheep to graze on the common, that will result in the greatest overall number of lambchops. That optimum number of sheep turns out to be about six. If we compute the self-interest of our individual farmer now, let's assume he has one of the six already. The farmer's sheep is worth 0.5 lambchops at this sheep density. Our farmer buys one more, and in the end has two sheep that will each produce 0.4 lambchops, for a total of 0.8 lambchops. It turns out to be in the farmer's self-interest to add sheep, up until there are nine sheep on the yard.

The point is that each farmer's personal best interest is served by grazing as many sheep as he can afford, in spite of the damage to the field. This results in a tendency to overgraze the field. From each farmer's point of view, the other villagers are all also overgrazing the common yard, so his own personal sacrifice, to preserve the common, would be in vain anyhow, and would put his family at a disadvantage. The end result is far more sheep than can be most efficiently supported on the common, decreasing the net yield of sheep in total. Everyone might be better off if they backed off a little bit and cooperated.

There are two potential approaches to preserving the common space. One is to divide it up and assign property rights to individuals. In this case, each individual would have the incentive to preserve the field by not overgrazing because that individual would directly benefit from careful stewardship.

However, for our example of land stewardship, private ownership is not a universal environmental panacea because some element of commonality remains. The aquifer holding water underneath the ground may be overtapped, or soils may be depleted throughout the lifetime of the individual (a sort of commons over time). For fisheries, the ownership society system has holes in it because fish swim around. For the case of global warming, dividing the atmosphere up into parcels is impossible by the nature of the problem. The other potential solution is for some form of collective self-regulation (see Negotiations section).

The decisions made by a single individual are often irrational, and therefore not really well described by economic theory. The collective decisions made by whole societies of individuals, in contrast, often exhibit great sagacity. Economics aficionados speak with reverence of the ***market***, economist Adam Smith's "invisible hand." The market with its law of supply and demand is a negative feedback system, just like the many natural examples in Chapter 7.

However, the market has its blind spots, leading to effects like the tragedy of the commons. In economics, a cost that is not paid by the decision maker is called an ***external*** cost. An example of an external cost associated with driving to work is the traffic. One more car will tend to slow down all the other cars on the road, costing other drivers their time. The cost of climate change is not only paid by people who are responsible, but by everybody, soon and far into the future. Our sheep farmer made the other farmers pay part of the cost of his new sheep. If true costs are left external, then the economic balancing act of the market does not take them into account, and tragedies of the commons are the result. An external cost can be ***internalized*** by means of taxes or regulations. The idea is to make the market aware of the true cost of a decision to take this path versus that path.

One way to harness the balancing abilities of the market for preventing global warming is a scheme called cap and trade. A regulatory agency allocates permits for emission of climate forcing agents such as CO_2, with the total number of permits they issue totaling some lower overall rate of emission than BAU (climate geek speak for business-as-usual). If a company or a country is able to cut its emissions to even lower than its allocation, it has the right to sell its permit to another company or country. The price of the permits is set by supply and demand, same as any other commodity. Presumably the industry for which it is most expensive to reduce emissions would be willing to pay the most for the emission permits. It will induce other industries to reduce their emissions because they can make money doing it, selling their permits to the highest bidder.

In this way, the market finds the most efficient, that is to say the cheapest, means of limiting global CO_2 emissions to the cap value. A cap-and-trade scheme in the US for sulfur emissions (generating acid rain) from power plants has apparently worked well. The difficulty with cap and trade for a regional problem like acid rain is that it may allow all the pollution to concentrate in one region, which may be too much for that region. CO_2 is a truly global problem, in contrast, and it makes no difference where the CO_2 is released, so CO_2 emissions would be an ideal problem for limitation by cap and trade.

The market may influence the global warming debate through the insurance industry. We saw in Chapter 12 the huge increase in insurance payout to cover atmospheric-related natural disasters. It is not at all clear to what extent this trend is due to climate or to social factors, but the insurance industry is certainly concerned about the possibility of climate change affecting their bottom line. Insurance companies have been responsible for other preventative measures such as requirements for hard hats on construction sites.

Pollution problems get more difficult to solve, politically, as their footprints expand from regional to global, because of the commons effect. Local issues like contaminated drinking water or urban smog are clearly in the interests of local individuals to solve. You fix it same as you would fix the roof over your head. A clear link exists between the costs of cleaning up and the benefits. Larger-scale, regional problems tend to run into us-versus-them issues. Why should the state of Ohio sacrifice its right to burn high-sulfur coal when the costs of that consumption are paid by the folks in New York? Why should power plants in Chicago pay to clean up their mercury emissions when most of the mercury emission to the atmosphere comes from China? At the most difficult, worst end of this problem, global warming is the most global of issues, the most complicated type to solve.

Global warming is also a difficult issue to address because it is long term. Values need to be related not only from one place to another, but also across time. The way that economists deal with values across time is via a construct known as a ***discount rate***. The idea is based on a question: which would an individual prefer, a cost today or a cost in the future? A cost in the future, certainly, would be the natural response. A rational reason to justify that choice is that one could invest money today and use the interest to help pay the costs in the future. If we were to assume an interest rate of say 3% per year, we could pay a cost of $100 in 100 years by investing $5 today (Fig. 13.2). It would be worth paying $4 today to avoid a $100 cost in 100 years, but it would not be worth paying $6 today. The bottom line of this idea is that costs tend to shrink with time.

Of course, the idea of anybody actually going out and opening a bank account with $5 in it, to save for that cost in 100 years, seems rather quaint. However, economic theory is more than just a prescription for how one could rationally make the most of one's assets; it is also a description of how money tends to flow in our society. The discount rate theory illustrates why financial decisions made by the market tend to be somewhat short-sighted. It's a description of the way money flows. Water flows downhill. Money flows short term.

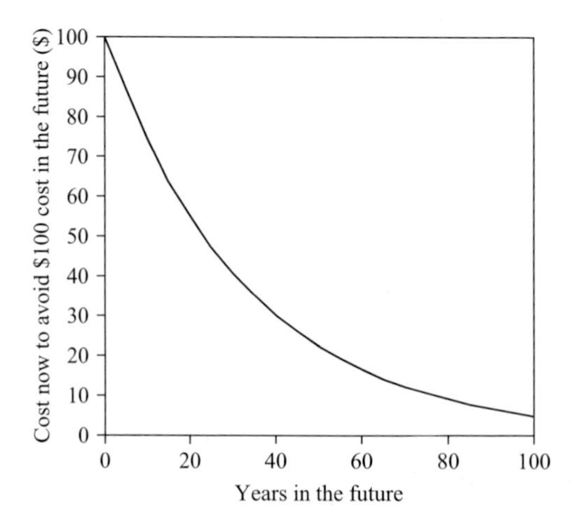

Fig. 13.2 **The discount rate and future costs. How much would it be worth, today, to avoid a cost of $100 at some point in the future, assuming a 3% discount rate?**

Negotiations

In 1988, agencies of the United Nations formed the IPCC, consisting in general of government scientists. The role of IPCC is to assess "the scientific, technical and socio-economic information relevant to understanding the scientific basis of risk of human-induced climate change, its potential impacts and options for adaptation and mitigation."

The purpose of IPCC is to publish reports summarizing the scientific literature. These reports are divided into three topic areas, written by different working groups (WG). WG I deals with the scientific basis for the climate change forecast. WG II deals with the impacts of climate change on the natural and human world. WG III assesses options for limiting greenhouse gas emissions or otherwise avoiding climate change. The most recent reports from all three are available online at http://www.ipcc.ch/. The next report will be released in 2007.

The reports consist of chapters and various levels of summaries. Much of the work of writing and reviewing the documents is done by researchers at universities, not necessarily employed by any government. The rules are that the chapters may present no previously unpublished research, but are based only on published, peer-reviewed scientific papers.

Of course, not every paper published in the scientific literature is guaranteed to be correct in every way, but the practice of sending manuscripts out to rival scientists for review makes sure that most of them are at least plausible. If a scientist disagrees with a paper that has already been published, he or she is free to write another paper attempting to disprove the first. Papers are sometimes retracted by their authors if they are proven wrong, or sometimes a disagreement just festers in the literature, requiring judgment on the part of the readership to decide which is right and wrong.

Once an IPCC chapter is drafted, the chapter is subject to another round of reviews by scientific experts and government workers. The chapters are grouped together into books which also come with technical summaries and summaries for policymakers. These summaries are subjected to line-by-line review, voting, and approval by the entire IPCC. This sounds like a long and painful process!

The conclusion of the first, 1990, IPCC report was that it was not yet possible to detect human-induced global warming. This was largely because of the mismatch between CO_2-only model results and the thermometer record (Fig. 11.11b). By 1995 it had been worked out that the warm temperatures between 1900 and about 1940 could be attributed to increased solar intensity, whereas warming from 1970 to the present was clearly due to rising CO_2 levels (Fig. 11.11c). The 1995 report issued a now famous statement that "the balance of evidence suggests a discernable human influence on global climate," thus providing the impetus for drafting the Kyoto Protocol in 1997.

Negotiations to limit CO_2 emission began with a document called the ***Framework Convention on Climate Change*** or ***FCCC*** that was drafted during the Earth Summit meeting in Rio de Janeiro in 1992. The objective of the FCCC was to achieve "stabilization of greenhouse gas concentrations in the atmosphere at a level that would prevent dangerous anthropogenic interference with the climate system." FCCC deferred defining what the dangerous concentration would be, a discussion that continues to this day.

FCCC established a procedure for negotiations by setting up a series of meetings between countries, called ***Conference of Parties*** or ***COP*** meetings. FCCC also established a procedure for creating amendments to itself, the most famous and significant of which was drafted during the COP-3 meeting in 1997 in Kyoto, Japan, called the ***Kyoto Protocol***.

The Kyoto Protocol attempts to limit global CO_2 emissions to about 6% below 1990 emission levels, by the year 2010. The countries participating in the negotiations (the "parties") are divided into industrialized and developing nations. The agreement limits the emissions of the industrialized nations, but does not put limits on emissions from the developing world. The agreement allows for cap-and-trade arrangements, allowing countries to buy and sell emission credits. Developed countries are allowed to invest in carbon-reduction measures in developing countries, and get carbon emission credit for this. Carbon uptake by reforestation is allowed, and counts toward meeting emission targets. This is some discussion of penalties for exceeding emissions, including a factor of 1.3 "make-up fee" for emissions that exceed the targets. However, the agreement has no real teeth; it remains to be negotiated whether the consequences for failure are truly legally binding.

The treaty came into force when countries accounting for 55% of CO_2 emissions globally agree to its terms. The United States withdrew from negotiations under the Bush administration in 2001, presenting a fairly sizable obstacle to achieving this target, seeing that the United States accounts for 25% of CO_2 emissions globally. In spite of this setback, the agreement came into force when Russia agreed to the Kyoto Protocol in 2005.

The Kyoto Protocol is only the first step toward a solution to the climate change problem. The target CO_2 emission rate from Kyoto is about 6% below 1990 levels. You may recall from Chapter 10 that the uptake rate by the atmosphere and ocean combined is about 4 Gton C/year, about 40% below 1990 levels. The actual projected climate impact of the Kyoto Protocol by itself is discouragingly small because of the small target cut, and because there are no limits to emissions by developing countries.

A major objection raised by the two "dropouts" from the Kyoto Protocol, the United States and Australia, is the exemption of the developing world from CO_2 emission limitations. The Kyoto Protocol has taken the approach of allocating acceptable CO_2 emissions for each country in the industrialized world according to how much CO_2 they currently emit. By far the greatest fraction of the carbon emissions, historically and today, comes from the industrial world. Capping future emissions according to historical emissions for the developing countries would therefore carve into law the imbalance between industrialized and developing nations.

We have seen in Chapter 10 the intimate relationships among CO_2 emission, energy, and economic activity. Within our current technological stage of development, CO_2 emission is the key to national wealth and comfort. The developing world understandably wants the benefits of CO_2 emissions also, and can claim that capping their emissions according to current emissions would be unfair. This is the rationale for exempting developing nations from emissions caps for the time being. In this strategy the Kyoto Protocol follows the example of the Montreal Protocol banning chlorofluorocarbon emission, which initially applied only to developed nations and then phased in to restrictions for everyone. Ultimately, because of the large numbers of people in the developing world, in particular India and China, the developing world is projected to account for a much larger fraction of CO_2 emission in the future. For this reason, it will be essential for continued negotiations beyond Kyoto to bring the developing world on board.

Economics

One way to approach the decision of whether to limit CO_2 emissions is based on economics. Which path costs less, to prevent climate change aggressively or rather to endure it? A few caveats are in order here before we proceed. Economic behavior is in general not as easy to forecast as climate can be forecast. Physical sciences are based on a solid foundation of natural laws such as Newton's laws of motion, conservation of energy, and so on. It is not always possible to derive the physics of climate solely from these natural laws. Higher-level processes like turbulence and clouds are called "emergent behavior," surprisingly difficult to build from the fundamental laws of physics. But at least the foundation exists. Economics and the other social sciences are fuzzier. Models of economic trends tend to be empirical, eyeball curve fits. It is not that physical scientists are smarter than social scientists, it is simply that physical sciences are more tractable.

Economic trends are strongly impacted by technological progress, which is impossible to forecast, and by social and political fancies. One could even argue that economics is pathologically unforecastable because it is self-aware. If I came up with a model that could successfully forecast stock prices, I could use it to make money, until everyone else discovered my methods. At that point, the stock prices would reflect everyone's knowledge of the model forecast, and the future trajectories of the stock prices would alter to evade my forecast. I wouldn't make money any more. Not even on my darkest days do I imagine that the physical world that I model for a living is going to change its behavior for the purpose of making my model wrong!

In general the costs of climate change are predicted to be nonlinear with temperature increase. A small increase in temperature, say 1°C or less, will probably not do too much damage, and may even be beneficial on the global average. As the temperature rises to 2°C or 3°C or higher, however, the economic impact starts to rise sharply. This is true for many aspects of climate change: farming, sea level rise, health impacts, and the possibility of abrupt climate change. William Nordhaus, an economist at Yale, has constructed an economic model to gauge the costs of climate change versus the costs of decreasing CO_2 emissions.

Economic costs are presented as a percentage in global domestic product, ***GDP***. Let's imagine an ongoing cost to the economy of 3% (Fig. 13.3). The cost is like a sales tax, applied to all the money that we make, every year. At the same time, the economy is growing at some rate. A typical growth rate might be 3% per year.

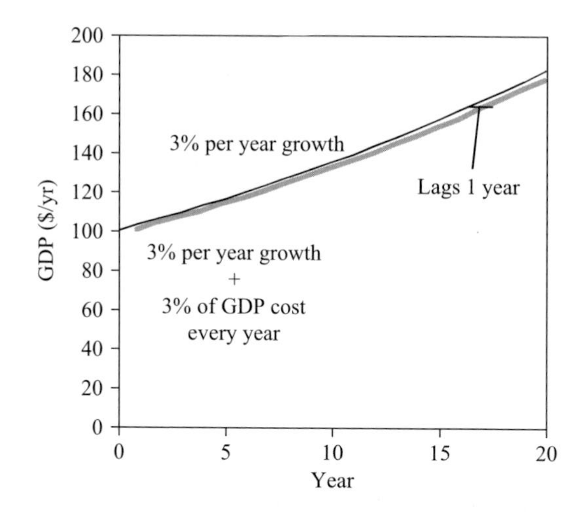

Fig. 13.3 **The effect of an ongoing expense equivalent to 3% of GDP every year, on an economy which is growing at a rate of 3% per year.**

This sounds small, a few pennies out of a dollar, but yet somehow people's livelihood and sense of prosperity depend extremely sensitively on that few percentage growth rate. If the economy grows by 5% each year, everyone gets rich and shoeshine boys trade stock tips. The great depression decreased global productivity by about 25%. The reunification of Germany put a drain of about 5% on the GDP of West Germany, producing noticeable malaise among West Germans.

In Fig. 13.3 we put together the 3% cost and the 3% per year growth rate in GDP. We see that the effect of the cost is to set everything back a year. A person would only get $97 this year instead of $100 because of that 3% cost. But by next year, the economy will have grown to $103, and she'll get $100, just one year later.

A 3% cost affects the growth rate of the GDP by only 3%, changing, say, a growth rate of 3% per year to 2.91% per year. In the long run, this does not seem like a long-term catastrophe to a naïve noneconomist like your author. To an Earth scientist, any positive exponential growth like this is an explosion, whether it explodes quickly or slowly; it cannot persist forever.

Costs of climate change

The projected agricultural costs seem rather small, perhaps 1% of GDP, because agriculture itself is only 2.5% of GDP or so, globally. Of course, this is a special 2.5%, different from say shoes or music videos because one cannot eat shoes or music videos.

The costs of sea level rise are projected to be of the order of 0.1% of GDP. Storm surges currently cost about this much every year. It is not unreasonable to imagine doubling this cost by increasing sea level, just to get an idea of the scale of how big it could be.

Health impacts are projected to be worth 0.5% of GDP globally to avoid, with higher impacts in tropical developing countries in Africa and Asia, where it can reach 3%. Of course, this statistic requires coming up with a way to express the worth of a human life in dollars.

The largest impact in Nordhaus' tabulation was a willingness to pay to avoid catastrophic climate change, which he defines as equivalent to the Great Depression, a 25% reduction in GDP, that lasts forever. He gauged the probability of an event like this by surveying experts in climate science, to derive the collective opinion of the group that there is a 1% probability of catastrophe if the globe warms by 2.5°C. If climate warms to 6°C, the probability, pulled out of the air by the panel of experts, rises to about 7%. When the probabilities and downside costs are combined in a way that recognizes that people are willing to pay a little extra in order to reduce risk, Nordhaus derives that it would be worth about paying about 0.5% of GDP globally as insurance to prevent rise in temperature by 2.5°C, or about 5% of GDP to prevent a temperature rise of 6°C, on the basis of catastrophic climate change alone. Overall, Nordhaus estimates that it would be worth paying 1.5–2% of GDP to avoid a 2.5°C rise.

Costs of Kyoto

The costs of climate change on the one hand can be compared with projections of the cost of compliance with Kyoto on the other. Of course these numbers are speculative as well. The costs of limiting CO_2 emission will depend on technology development, very difficult to forecast. Ultimately only time will tell. These are guesses, but it is interesting to look at the numbers anyway.

The forecast is that compliance for Kyoto by all nations including the US would cost about 1% of GDP each year, according to a book published in 2000, and 7% in a paper in Science published in 2002. The Science article is the more recent of the two, and the cost went up apparently because the predicted BAU baseline had gotten higher. (A change in our expectations appears as a cost.) Interestingly, the cost to the United States is much higher than the cost to other countries because the carbon emissions are rising much more slowly in other developed countries than it is in the States.

US participation would be a boon to the rest of the industrialized world because we have such a ravenous appetite for fossil fuels that we would be forced to purchase CO_2 emission credits from other countries, so that the price of those credits would be much higher. Nordhaus' model forecasts that, without US participation, CO_2 emission credits would trade for about $40 per ton of CO_2 in the year 2025. United States' participation would drive the price up to $150 per ton of CO_2.

The bottom line is that compliance with Kyoto could cost a few percentage of GDP every year, and climate change could cost a few percentage of GDP every year. Don't forget that Kyoto will not prevent global warming, it is only the first step, and further restrictions will cost more. One could therefore uncharitably characterize Kyoto as "inefficient and expensive."

It may still however be better than the alternative. The costs of climate change will persist as long as CO_2 persists, say 300 years for the bulk of it. If you compute the total costs added up over the long term, Kyoto seems like a bargain because we pay for a few decades and benefit for hundreds of years. Ah, but don't forget that discount rate. We don't care about costs or benefits in the future; we want the money now.

Economics is an awkward tool for the climate change decision because many aspects of the natural world are simply not economically fungible. The economic value of the coral reefs could be tabulated in terms of the tourist dollars they generate, same as Euro-Disney, but somehow something is missing in this tabulation. If coral reefs cease to exist because it gets too hot, we cannot purchase more coral reefs for any amount of money.

The bulk economic framework also hides an issue of fairness, in that people who benefit from BAU are not the same people who pay for BAU, say by living in an increasingly drought-prone world far in the future. I imagine that giving up the institution of slavery in the American South entailed a cost, too, for somebody. Ultimately the abolition of slavery was not an economic decision, but an ethical one.

Although the Kyoto Protocol will not by itself prevent climate change, it would provide the benefit of guiding investment toward decreasing CO_2 emission. Coal power plants, for example, can be built using newer technology called "integrated gasification combined cycle" (IGCC), enabling future CO_2 capture for sequestration into geological repositories. If CO_2 is released into the atmosphere, IGCC plants cost about 20% more than traditional plants. If CO_2 sequestration starts to be common practice, however, IGCC plants would be cheaper overall than traditional plants. In the regulated US energy market, regulations often require power companies to use the cheapest technologies. In general this regulation makes sense, if we want to limit the power of monopoly power companies to spend money wantonly, passing the costs on to their captive customers. But in this case, the cost of climate change is external to the accounting, and so the regulation may not be choosing the cheapest option. Once a power plant is built, it has a life span of 50 years or so. The ultimate costs of decreasing CO_2 emissions would be lowered by starting now, rather than continuing to invest in high-emission technology.

Dangerous anthropogenic interference

Another way of approaching the problem is to ask, What would it take to avoid dangerous climate change? This was how the FCCC was constructed, to "prevent dangerous anthropogenic interference with the climate system." IPCC didn't specify what dangerous interference is, leaving that topic to future discussion which is still happening.

One way to define dangerous interference is to base it on the range of climate variability that the Earth has experienced in the last million years or so. If we change our climate to that of the age of dinosaurs, that sounds dangerous. Let's not do that. That's the idea. Global mean temperature may have been 1.5°C warmer during the last interglacial period, 120 kyear ago, or the last time the Earth's orbit resembled the way it is today, a time called stage eleven, 400 kyear ago. Based on that, a scientific panel advising the German government called the German Advisory Council on Global Change (http://www.wbgu.de) argues that 2°C is a reasonable maximum tolerable global increase in temperature.

Another way of defining dangerous is to watch sea level. Jim Hansen (2004) argues that 2 m of sea level rise sounds dangerous to him. The IPCC forecast for sea level rise is about half a meter in the next century, detrimental but not yet dangerous according to Hansen. But 2 m, you're into some serious Florida. That sounds like a big deal. The longer-term trends from the past (Fig. 12.6) seem to predict tens of meters of sea level change with long-term doubling of CO_2. The question is how quickly the ice melts down. Heinrich events (Chapter 12) suggest that ice can melt pretty quickly if it has a mind to, although we're not too clear on how it manages this. Hansen argues that the past would suggest the possibility that 1°C temperature increase beyond today might exceed his 2 m sea level danger limit.

The other issue is how fast the temperature change occurs. FCCC says that warming should be slow enough for ecosystems to adapt naturally, with no threat to food production, "and to enable economic development to proceed in a sustainable manner." In other words, the transition to a new climate regime should be slow enough for natural and human systems to adapt. Here again history can be our guide. The trick is, short warming spikes heat up more quickly than long-term trends, so we have to compare temperature trends computed over similar lengths of time. The current rate of warming globally over the past decade is 0.2°C per decade; higher than this in some locations regionally. When we compute the rate of change over multiple decades, the recent rate drops to 0.1°C per decade on average. Historically, warming at 0.1°C per decade is a fairly unusual warming. Based on this, the German Advisory Council (http://www.wbgu.de) concludes that 0.2°C per decade seems about the upper safe limit to the long-term temperature rise.

Now we need to translate from units of degree centigrade that we consider dangerous to atmospheric CO_2 concentration that would give us that temperature, then to CO_2 emissions to stay within that atmospheric CO_2, then finally to energy policy to keep within those carbon emissions. Beginning with a target temperature that we consider to be dangerous, we use ΔT_{2x} to calculate the atmospheric pCO_2 value you need to get there (Chapter 4). According to Eq (4.1), a temperature change of 1°C is already in the pike from our pCO_2 concentration of 360 ppm; 2°C would be generated by pCO_2 of about 450 ppm; 560 ppm would generate about 3°C (Fig. 4.7).

Translating from atmospheric CO_2 concentration back to the CO_2 emissions that would have generated that concentration requires a model of the uptake of CO_2 into the ocean and the terrestrial biosphere, such as the ISAM carbon cycle model (Chapter 9). In order to compare the responses of different models, they must each be subjected to the same trajectory of atmospheric CO_2 rise because the amount of CO_2 the ocean can absorb depends on how much it has already absorbed (its history). IPCC in 1992 constructed for this purpose a series of what-if scenarios for the trajectory of atmospheric CO_2 through the year 2100, called CO_2 stabilization scenarios. These are arbitrary but reasonable looking trajectories for the CO_2 rise in the atmosphere, ending up at a stable concentration of 350, 450, 550, 650, or 750 ppm.

The online model for the Kaya Identity (http://understandingtheforecast.org/Projects/kaya.html) has taken advantage of the CO_2 stabilization scenarios to estimate how much carbon-based energy must be replaced by some new form of carbon-free energy, if we are to achieve atmospheric stabilization. Recall from Chapter 9 that the Kaya Identity model makes a forecast of energy use and carbon emissions to the year 2100 as a function of population size, economic growth, energy efficiency, and energy sources. Then we compare the carbon emission rate from the Kaya model, say for the year 2100 in Fig. 9.14 it is about 700 ppm. The ISAM model tells us that about 2.5 Gton C can be emitted in 2100 for the CO_2 to follow the stabilization scenario for 450 ppm. Kaya wants 18 Gton C/year, but to stabilize at 450 ppm, we can only emit 2.5 Gton C/year, so we will have to cut emissions by 15.5 Gton C/year.

Coal could generate 17 TW of energy from this carbon emission, so to replace coal we need about 17 TW of carbon-free energy (Fig. 9.14). Coal would be the most sensible carbon fuel to replace, and it may be all the fossil carbon that's left toward the end of the century, if we don't start digging into clathrates or oil shales. How much is 17 TW of energy? For comparison, mankind is currently consuming about 13 TW of energy. The sun bathes the Earth in energy at a rate of about 173,000 TW. Photosynthesis captures about 100 TW.

Alternatives

One obvious possibility is conservation. The United States uses twice the energy per person as is used in Europe or Japan (Fig. 9.11). In part this is because the human terrain of the United States developed in a more automotive way, whereas public transit and trains are better developed in Europe and Japan. In addition, Americans drive larger cars and live in larger houses. I personally enjoy life in Europe. I would far prefer to take a train to work, in which I can read or watch people, rather than curse at other drivers in a private car. Certainly it makes more sense, in a global sense, for Americans to learn to live like Europeans rather than Europeans and other earthlings to learn to live like Americans.

On the other hand, it seems unlikely that conservation alone will solve the world's energy problems. Most of the world lives at a standard far below that of Europe or the United States. If the citizens of the developing world ultimately consume as much as the European model, the footprint of humanity on the Earth would increase by about a factor of five. The Kaya projection of energy use in the future already assumes increases in energy and carbon efficiency, and it still predicts a thirst for energy that results in higher CO_2 concentrations than the stabilization scenarios.

None of our traditional sources of carbon-free energy seems capable of making up the shortfall for the coming century by itself (Hoffert et al. 1998; Pacala and Socolow 2004). The first option that seems to come to people's minds is ***nuclear energy*** because it generates no carbon, and we know it works. The French are using nuclear for more than a third of their energy needs. However if we want to generate 17 TW of nuclear energy, using present-day standard 1000 MW reactors, it would require 17,000 new reactors within 100 years, for an average reactor construction rate of one new reactor every second day, continuously, for 100 years. This seems like a lot of nuclear reactors, or else development of reactors much larger than today's.

It is a close call whether or not there would be enough minable uranium to support this level of nuclear energy production. The global inventory of minable uranium today would be sufficient to generate 10 TW of energy at our current efficiency for a few decades before it would be exhausted. However, uranium today is rather inexpensive, and we only bother to mine the purest uranium deposits because production of nuclear energy is limited by supply of reactors, not by supply of uranium. We could move to lower-purity uranium deposits. Ultimately it may be possible to extract uranium from seawater, where it is present in huge amounts but at very low concentrations. In addition, it is theoretically possible to stretch the energy yield of natural uranium by converting it to plutonium in what is known as a ***breeder nuclear reactor***. This increases the energy yield from the uranium by a factor of 50, but the downside is that it is very easy to produce nuclear weapons from plutonium. For this reason, breeder nuclear reactors are not used today.

Opposition to nuclear energy arises because of the potential for catastrophic accidents. The explosion at the Chernobyl power plant in the former USSR could have been much worse if the core had melted down to the water table, contaminating the groundwater supply to Ukrainian agriculture and the city of Kiev. Nuclear energy has the potential to poison large stretches of land to human use essentially forever. The other objection to nuclear energy is the problem of waste storage. Breeder reactors have the advantage of incinerating the long-lived radioactive wastes.

Windmills are becoming economically competitive with new traditional power plants. (The trick word in that sentence was "new." An already existing coal power plant can be run very inexpensively; nothing can compete with that.) Windmills supply 7% of the energy needs of Denmark with no adverse impact on the beauty of the landscape of Denmark in my opinion. Wind energy currently accounts for 0.3% of energy globally and is growing at 30% per year. At this rate wind could supply 10% of the world's energy within the next couple of decades. Scaling up current wind energy production by a factor of 50 would generate about 2 TW of energy.

It may be possible to extract energy from winds at high elevations in the atmosphere. Winds get faster with altitude in the troposphere, peaking in the jet stream winds in mid-latitudes of both hemispheres. The air up there is at lower pressure than down here, and therefore has less density to drive wind blades, but still the power density is much higher at high altitude than it is at the Earth's surface (Fig. 13.4). Wind energy collectors could function like giant kites tethered to the ground by electrically conducting cable, remaining aloft passively by riding the wind. It has been proposed that high-altitude windmills could scale up to tens of terawatts of energy (http://www.skywindpower.com).

Solar cells, or ***photovoltaics***, generate electricity from sunlight. Photovoltaics are currently rather expensive to produce, but if solar energy generation were scaled up by a factor of 700 from present-day rates, they could generate several terawatts of energy. Similarly to surface windmills, photovoltaics seem unlikely to solve the entire energy problem but they can certainly help. One new idea is to build solar cells on the moon (Hoffert et al. 2002), beaming energy back to the Earth as microwaves. The moon would be an ideal location for solar cells, in that it never rains there, there are no clouds, and no birds. The cells could be produced from lunar soil, reducing the amount of mass that would have to be lifted out of Earth's gravity by rockets, a huge savings

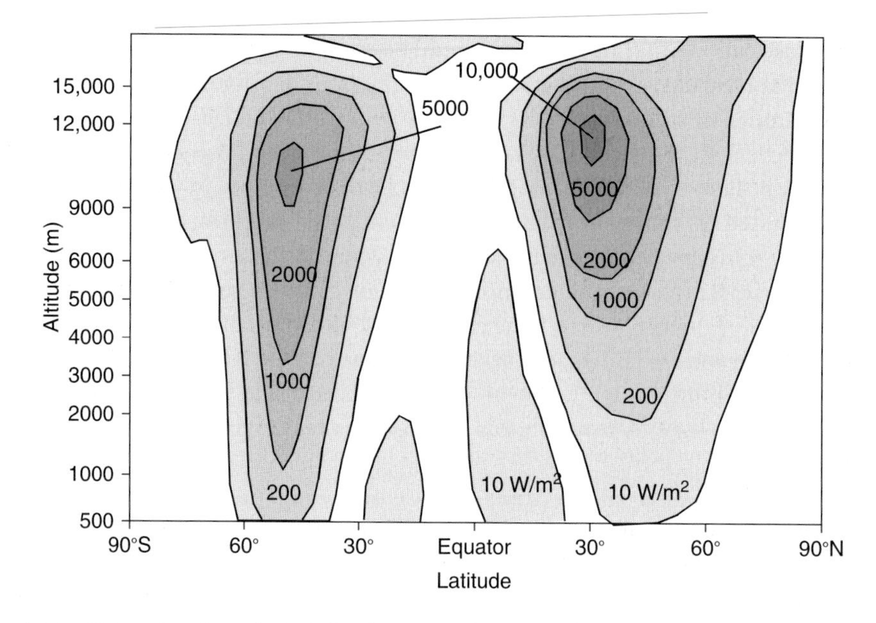

Fig. 13.4 **Power density of winds in the atmosphere as a function of latitude and altitude. From Caldeira, via windskypower.com.**

in energy and cost. The energy from the solar cells could be transmitted to Earth via microwaves; this is apparently feasible, safe, and relatively efficient. An array of solar cells in orbit around the earth would also share the benefits of a clean environment and solar flux that is unimpeded by the atmosphere. Solar cells in orbit however have to be produced on Earth and lifted to orbit.

Both surface wind and solar power share a technical difficulty that energy must be stored because periods of maximum power demand may not coincide with times of maximum windiness or sunshine. To some extent, each technology benefits from the other, in that, when the sun ain't shining, the wind often be blowing. Another way to deal with this problem is to globalize the power grid, so that electricity could be efficiently transported around the world. Electrical energy can also be used to generate ***hydrogen***. You have probably heard the hype about the hydrogen economy. Hydrogen itself is not a primary energy source. There are no minable sources of energy as hydrogen like the energy in fossil fuel deposits. Hydrogen can be produced from other sources of energy, for example, electricity can be used to break water into components hydrogen and oxygen gas, a process called ***hydrolysis***, or produced chemically from coal in a process called ***gasification***.

Hydrogen could be used to store energy for use in transportation. Because hydrogen is a gas, it must be contained under pressure, as opposed to gasoline, which is liquid at room temperatures, or propane, which can be liquefied at moderate pressure. Propane is already being used to power some buses, as well as things like forklifts. Hydrogen is a very flammable gas, and burns invisibly unless the flames happen to hit some other material that emits light. Hydrogen is more explosive than propane, but in crash tests, the added danger is offset by the tendency of hydrogen to rise up, like helium would, escaping from a crashed vehicle.

Another option for carbon-free power is to use fossil fuels and then dispose of the carbon, called ***CO_2 sequestration***. CO_2 could be captured from the exhaust stream of a power plant, or coal could be treated with steam to release hydrogen gas and CO_2. The CO_2 could be injected into the Earth into geological formations that must be porous enough for CO_2 to flow away from the injection site, but isolated from the surface so that the CO_2 doesn't escape back to the surface. The largest type of geological formation that would fit the bill is called saline aquifers. These deposits contain water in their pore spaces, but the water has salt dissolved in it, so the assumption is that this water is no good for anybody and we might as well inject CO_2 into it. Methane gas has remained stable in deep Earth reservoirs for hundreds of millions of years, so the idea is in principle possible. These aquifers are thought to have the capacity to store 10,000 Gton of C as CO_2.

Scientists are also discussing the possibility of sequestering CO_2 in the deep ocean. CO_2 released into the atmosphere will ultimately mostly dissolve in the ocean anyway; 75% of the carbon is in the ocean when the air and water reach equilibrium after hundreds of years. One could envision direct injection into the ocean as simply bypassing the transit of CO_2 through the atmosphere. CO_2 released in the ocean will equilibrate with the atmosphere toward the same equilibrium point; 25% of ocean-released CO_2 would escape to the atmosphere after hundreds of years. The immediate difficulty with ocean injection is that the CO_2 is very concentrated right near the injection site, acidifying the water (Chapter 10) and killing marine life.

You may have read about the idea of fertilizing the plankton in the ocean to take up CO_2 as a means of carbon sequestration. The idea is superficially attractive in that plankton in many parts of the ocean, in particular in the cold surface waters around Antarctica, are starving for tiny amounts of the element iron. One atom of iron would allow plankton to take up 100,000 atoms of carbon (more or less). Iron is typically supplied to the surface ocean through dust blown through the atmosphere, but the Southern Ocean is remote enough that not much dust blows down there. When oceanographers add iron to surface water, the plankton bloom. The idea is that the plankton would grow, die, then sink to the deep ocean, thus taking charge of the jobs of extracting CO_2 from the surface ocean, ultimately from the atmosphere, and transporting it to the deep ocean for us. The problem is that models of the carbon cycle in the ocean and atmosphere predict that even if the entire Southern Ocean could be successfully fertilized, the effect on atmospheric CO_2 in the coming century would be discouragingly small. The reason is that it takes hundreds of years for the atmosphere and ocean to reach equilibrium. Ocean fertilization could have a larger impact on atmospheric CO_2 if we were willing to wait 500 years. The impact in the next few decades is too small to be worth it.

The last possibility to mention is the idea of deliberately altering the climate in such a way so as to counteract the warming effects of rising CO_2 concentrations. One possibility is to deliberately inject sulfate aerosols into the stratosphere. Sulfate aerosols cool by scattering light. Particles in the stratosphere remain there for several years, as opposed to aerosols in the troposphere where rain scrubs them out in a few weeks. The cooling effect of volcanic eruptions such as Mt. Pinatubo has been clearly documented. The aerosols could be shot up to the upper atmosphere in 1-ton shells fired by large cannons. Relative to most of the alternatives, cooling the Earth by this method seems inexpensive and benign. My own personal objection to this idea is that the warming effects from CO_2 will last for centuries, even millennia, committing mankind to taking an active role in climate maintenance essentially forever for a few decades of carelessness. Other proposals for climate engineering include the placement of a large reflective object in space, in orbit around the Earth or at the stable "Lagrange" point between the earth and the Sun where objects can sit indefinitely. The space-based option would be more costly and would require active removal if its effects were eventually judged to be detrimental, but it may not require ongoing participation in order to continue working.

Summary

In summary, we appear to have multiple options for dealing with climate change within the context of continued economic growth, but none of the options is trivial or without obstacles. Socially, politically, and technologically, the issue of climate change poses a challenge to mankind on a larger scale than humankind has ever had to face before.

Take-home points

1. Human-induced climate change is an example of the tragedy of the commons effect. Economists refer to climate change as an external cost, which can cause the market to make poor decisions. Economic forces also keep the market focused on short-term profit and cost, disregarding costs that come far in the future.
2. International negotiations under the auspices of the United Nations have resulted in a treaty called the Kyoto Protocol to limit CO_2 emissions. The emissions cuts mandated by the Kyoto Protocol are small compared with what would be needed to avoid climate change. The treaty also only imposes cuts on industrial countries, who are currently emitting the most CO_2. For these reasons, the Kyoto Protocol can only be the first step toward preventing anthropogenic climate change.
3. The ultimate goal of the negotiations is to prevent "dangerous interference" with the climate system. The definition of "dangerous" within this context might be in terms of temperature, or in terms of sea level change.

Projects

1. *Compound interest.* The formula to compute compound interest for a bank account is

$$\text{Balance}(t) = \text{Balance}(\text{initial}) \cdot e^{k \cdot t}$$

This was first presented in Chapter 5 when we met the exponential function e^x. Assuming an interest rate of 3% per year, how much would an investment of $1.00 made today be worth in the year 2100? What if the interest rate were 5% per year?

2. ΔT_{2x}. The formula to estimate temperature response to changing CO_2 concentration is

$$\Delta T = \Delta T_{2x} \times \frac{\ln(\text{new } pCO_2/\text{orig} \cdot pCO_2}{\ln(2)}$$

This formula was first introduced in Chapter 4. Calculate the temperature that would result from increasing atmospheric pCO_2 from preanthropogenic value of 280 ppm to 1000 and 2000 ppm. The temperature during the Cretaceous period might have been 6°C warmer than today. Using a ΔT_{2x} value of 3°C, what must the pCO_2 have been at that time? How does your answer change if ΔT_{2x} is 4°C?

3. *Carbon-free energy.* The Kaya Identity web page actually runs a carbon cycle model to predict the atmospheric pCO_2 response to its predicted carbon emissions. You learned about this model in Chapter 9. The Kaya web page then computes how much coal energy would have to be replaced by carbon-free energy, if we wish to stabilize atmospheric CO_2 at some concentration (choices are 350, 450, 550, 650, and 750 ppm). Using the default web page settings, which are something like a BAU scenario, find from the plot the amount of energy in terawatts required to stabilize CO_2 at 450 ppm.

a. If a typical nuclear reactor generates 1000 MW of energy, how many power plants would be required by the year 2100? (The prefix tera means 10^{12}, whereas mega means 10^6.) How many power plants would this require?
b. A modern windmill generates about 1 MW of energy; let's say that future ones will generate 10 MW/tower. How many of these would be required to meet our energy needs by 2100? The radius of the Earth is 6.4×10^6 m. What is its surface area? Land occupies about 30% of the surface of the Earth; what area of land is there? Let's assume that windmills could be placed at a density of four windmills per square kilometer. What fraction of the Earth's land surface would be required to supply this much wind energy?

Further reading

Climate protection strategies for the 21st century: Kyoto and beyond, German Advisory Council on Global Change (WGBU), Special report, 2003.

Hansen, J., Defusing the global warming time bomb, *Sci. Am.* (2004), *290*, 68–77.

Hoffert, M.I., K. Caldeira, G. Benford, et al., Advanced technology paths to global climate stability: energy for a greenhouse planet, *Science* (2002), *298*, 981–7.

Hoffert, M.I., K. Caldeira, A.K. Jain, and E.F. Haites, Energy implications of future stabilization of atmospheric CO_2 content, *Nature* (1998), *395*, 881–4.

Nordhaus, W. and J. Boyer, *Warming the World: Economic Modeling of Global Warming*, MIT Press, Cambridge, 2000.

Nordhaus, W., Global warming economics, *Science* (2001), *294*, 1283–4.

Pacala, S. and R. Socolow, Stabilization wedges: solving the climate problem for the next 50 years with current technologies, *Science* (2004), *305*, 968–72.

Glossary

Absolute zero – the lowest possible temperature when atoms have the minimum possible amount of vibrational and translational energy. This is 0 K on the Kelvin scale, and −273.15°C on the centigrade scale.

Acid – a water solution containing high concentrations of H^+ ions.

Acid rain – the effect of sulfate aerosols, consisting of sulfuric acid, on the acidity of rainwater.

Active zone of permafrost – surface soil that melts in summer, with ice below.

Adiabatic – insulated from the surroundings, unable to gain or lose heat from the environment.

Albedo – the fraction of incoming solar radiation that is reflected back to space without ever being absorbed. Ice and clouds have a high albedo, so they cool the Earth.

Anaerobic – without oxygen.

Anthropogenic – arising because of people.

Atmospheric window – between 700 and 900 cm^{-1} (wave numbers), there are no greenhouse gases that absorb and emit infrared light. If there are no clouds, light emitted at the ground goes right out to space.

Band saturation – if there is a high enough concentration of a greenhouse gas, light within some frequency range that is absorbed by the gas will be completely absorbed. If the concentration of the gas is increased further, it will not absorb any more light of that frequency. The main absorption band of CO_2 is saturated, so adding one molecule of CO_2 has less impact on climate than one molecule of methane.

Benthic – living on the sea floor.

Blackbody radiation – emission of light from an object at some temperature greater than absolute zero. An object that is capable of absorbing and emitting all wavelengths of light is called a blackbody, and the spectrum of light that is emitted is called a blackbody spectrum.

Borehole records – the temperature in the interior of an ice sheet or in the ground reflects the temperature of the air today and also of the past. Borehole temperatures are one type of paleo-temperature proxy record.

Carbohydrates – carbon in the intermediate oxidation of building blocks of CH_2O.

Centigrade – metric temperature scale. 0°C is freezing, 100°C is the boiling point of water, −273.15°C is absolute zero or 0 K on the Kelvin scale.

Chaos – when small errors or differences grow with time. Since weather is chaotic, it is impossible to forecast weather for more than a few weeks into the future.

Cirrus clouds – wispy clouds consisting of ice particles in the upper troposphere.

Climate – an average year of the attributes of weather like temperature and rainfall taken over some years.

Climate forcing – some agent that can alter climate, such as greenhouse gas concentrations, solar intensity, volcanic dust, or anthropogenic aerosols.

Climate sensitivity – the amount of warming, averaged over the whole Earth and the whole year, that would ultimately result from doubling CO_2. This is abbreviated as ΔT_{2x}.

Cloud condensation nuclei – particles in the atmosphere around which water droplets can form. If condensation nuclei are scarce, the cloud will have fewer droplets, each of which is larger.

Sulfate aerosols act as cloud condensation nuclei leading to what is known as the indirect effect of sulfate aerosols.

CO_2 fertilization – stimulation of plant growth by rising CO_2 levels. Higher CO_2 levels may ameliorate water stress for the plants.

CO_2 sequestration – the idea of pumping CO_2 back down into the Earth or into the ocean, rather than emitting it to the atmosphere.

Conduction – transfer of heat through matter by physical contact.

Contrails – condensation trails left by aircraft flying in the upper troposphere visible as lines in the sky. After a few hours or days, these may be indistinguishable from cirrus clouds.

Convection – when a fluid is heated from below or cooled from above until there is dense fluid on top of less-dense fluid, so that it turns over.

Coral bleaching – corals under stress may expel the symbiotic algae that they harbor, leading to loss of color. This is often followed by death of the coral.

Coriolis acceleration – a fake force added to the equation of motion to make the equation work within a rotating frame of reference.

Cumulus clouds – tower-type clouds formed by localized upward air motion. These are found in thunderstorms and in the tropics.

Cyclonic – the direction that air flows around a region of low pressure, such as is found in a storm. In the northern hemisphere, cyclonic flow is clockwise when viewed from above.

Discount rate – also called an interest rate. In economic theory, a cost in the future is discounted, with the idea that one could invest a smaller amount of money today, and using the interest, pay off the cost in the future.

Dissolved inorganic carbon – abbreviated as DIC, this is CO_2, HCO_3^-, and CO_3^{2-}. There are about 48,000 Gton of C in the ocean DIC pool.

Eccentricity cycles – cycles in the circularity or elliptical character of Earth's orbit. Eccentricity varies on timescales of 100,000 and 400,000 years.

Ecological succession – the process by which an ecosystem changes through time, one group of species followed by another, until a climax ecosystem is reached.

El Niño – an oscillation of the structure of the atmosphere and ocean, centered in the equatorial Pacific but with climate impacts around the world.

Electromagnetic radiation – coupled oscillations of the electric and magnetic fields that together propagate through space. Visible light is electromagnetic radiation, as is infrared, ultraviolet, x-rays, and others.

Emissivity – written as ε (the Greek letter epsilon) a constant describing how good a blackbody an object is. If $\varepsilon = 1$, the object is a perfect blackbody.

Energy intensity – energy use divided by economic production. The amount of energy it takes to make a buck.

Ensemble – a collection of climate or weather model runs, beginning from slightly different initial conditions. When an ensemble of runs is averaged, the result is more reliable as a forecast.

Exponential function – based on the number e raised to a power x, that is, e^x. The exponential function grows or decays proportionally to the value at any time. Interest for a bank account, population growth, radioactive decay, and light absorption all follow the exponential function.

External cost – a cost associated with some action that is not paid by the decision maker. This leads to the tragedy of the commons.

Feedback – a loop of cause and effect. A positive feedback tends to amplify an initial change whereas a negative feedback stabilizes the system.

Foraminifera – single-celled protista in the ocean that secrete shells of $CaCO_3$, from which paleoclimatologists can produce paleo-temperature proxy records.
Framework convention of climate change – a procedure for international negotiations to limit CO_2 emissions.
Frequency – the number of somethings per time. The frequency of light is cycles of the electric or magnetic field per time. The frequency of hurricanes is the number per year.
Geopotential surface – a surface that is flat relative to local gravity field.
Geostrophic flow – the way that fluid in the atmosphere and ocean flows in equilibrium on the rotating Earth. Geostrophic flow is at right angles to the direction of forcing, from pressure or wind friction. In the northern hemisphere, water flows 90° to the right of the wind direction.
Gigaton – billion metric tons, equal to 10^{15} g.
Glaciers – ice rivers that flow down mountain valleys, originating as snowfall on the mountain.
Gradient – the amount of change in a quantity, such as pressure or a chemical concentration, per meter of distance.
Greenhouse gas – a gas which is capable of absorbing and emitting IR light, for example, CO_2, water vapor, methane, and ozone.
Hadley circulation – A planetary scale circulation consisting of rising air at the equator, poleward flow to 30° N and S, subsidence, then equatorward flow at the surface. The Hadley circulation generates a general pattern of rain at the equator, and deserts at 30° N and S.
Heinrich event – a massive iceberg discharge in the North Atlantic during glacial time, documented by deposition of rocks on the sea floor.
Humic acid – a form of organic carbon in soils, from the decay of living materials.
Hydrocarbon – carbon in the reduced form, chains with hydrogens.
Ice albedo feedback – warming leads to melting of ice, lowering of the albedo, leading to further warming.
Ice sheet – thick flowing ice that is grounded on land. Greenland and Antarctica are examples today.
Ice shelf – floating ice, sometimes hundreds of meters thick from an ice sheet that flows onto the water.
Ice stream – a region of the ice sheet where the ice flows much more quickly than the surrounding ice.
Infrared light – abbreviated as IR. Longer wavelength light than visible light. IR is emitted by objects at near room temperature.
Inorganic carbon – carbon that is fully oxidized. Types of oxidized carbon include CO_2, $CaCO_3$, HCO_3^- (bicarbonate ion), and CO_3^{2-} (carbonate ion) in seawater.
Intensity – used in this book to describe energy flow per unit area, W/m^2.
Intergovernmental Panel on Climate Change – abbreviated IPCC.
Isotopes – atoms that have the same chemistry but different masses. The composition of stable isotopes can be affected by temperature, so isotopic measurements can be used to create paleo-temperature proxy records.
Joule – a unit of energy, equal to 4.18 cal, where a cal is the amount of energy it takes to warm 1 g of water by 1°C.
Juvenile carbon – carbon that is released from the Earth's interior for the first time in the history of the Earth.
Kelvin – the absolute temperature scale. 0 K is absolute zero. 273.15 K is the freezing point of water, which is defined as 0°C on the centigrade scale. One degree kelvin equals one degree centigrade.
Kerogen – a form of organic carbon in soils, from the decay of living materials.

Kyoto Protocol – an international agreement to limit CO_2 emissions.
Lapse rate – the decrease in temperature of the atmosphere with altitude.
Last Glacial Maximum – 30,000–18,000 years ago, abbreviated LGM. Global mean temperature was perhaps 6°C colder during this time.
Latent heat – heat carried by water vapor, released when the water vapor condenses to liquid.
Little Ice Age – 1650–1800. A period of generally cool and rather unstable climate in Europe and North America, coincident with the Maunder Minimum.
Maunder minimum – 1650–1700. A period during which there were no sunspots, coincident with the Little Ice Age.
Medieval Optimum – about 800–1200 AD. A period of generally warm stable climate in Europe, coincident with a prolonged drought in the American southwest.
Metamorphic decarbonation – a chemical reaction wherein sedimentary rocks like $CaCO_3+SiO_2$ lose CO_2 to become a silicate rock like $CaSiO_3$.
Methane – CH_4, the main component of natural gas.
Moist convection – convection in the atmosphere with the added complication that water condenses into clouds when the air rises.
Montreal Protocol – the 1987 international agreement to curtail chlorofluorocarbon emission.
Natural gas – fossil fuel in a gas state, mostly methane.
Obliquity cycles – the angle of the Earth's axis of rotation, relative to the plane of the Earth's orbit.
Ocean ventilation – circulation carrying surface water into the deep ocean. Ventilated ocean water has higher concentrations of oxygen. Ventilation carries rising CO_2 into the deep ocean.
Organic carbon – carbon that is partially oxidized and partially reduced. Living carbon is generally in this state.
Oxidized – an atom that donates electrons to another atom in a chemical bond is oxidized. Oxygen is greedy for electrons, so atoms that are bound to it are oxidized.
Ozone – an molecule consisting of three oxygen atoms. Ozone in the stratosphere protects us from UV light whereas ozone near the ground causes respiratory problems.
Ozone hole – the observed severe depletion of ozone in the Antarctic stratosphere caused by an interaction of chlorofluorocarbons (freons) and stratospheric nitric acid ice clouds.
Parameterization – when a climate model is unable to simulate the real physics of some process, such as cloud droplet formation, the process is predicted based on some simplified "informed guess" type of model.
Permafrost – soil that has remained frozen for at least the last 2 years.
pH – the negative base 10 logarithm of the concentration of H^+ ions in a water solution. A pH value of 7 is neutral, less than 7 is acidic, and greater than 7 basic.
Photosynthesis – plants harvesting energy from the Sun and storing it as reduced carbon.
Planktonic – the word means free floating as opposed to the swimming "nekton" but in practice, plankton live in the surface ocean.
Precession cycles – the axis of rotation of the Earth spins relative to the elliptical dimensions of the Earth's orbit. Precession cycles dominate climate variability in low latitudes, and take 20,000 years.
Proxy records – paleoclimatologists extend their climate records into prehistoric time, using stand-in measurements such as tree ring thickness, because there were no thermometers back then.
Radiative convective equilibrium – a state of energy balance for each layer of air in the atmosphere when radiation and convection are both going on.

Radiative equilibrium – a state of energy balance for each layer of air in the atmosphere, when radiation is the only process carrying heat around.
Reduced – an atom that borrows electrons from another atom in a chemical bond. Hydrogen is loose with electrons, so atoms that are bound to it are reduced.
Relative humidity – the amount of water vapor in the air, compared with saturation. If relative humidity is greater than 100%, the vapor will tend to condense to liquid, until 100% is reached. If relative humidity is less that 100%, water will tend to evaporate.
Respiration – converting reduced carbon to metabolic energy by oxidizing the carbon to CO_2. Multicellular life uses oxygen for respiration, but bacteria can use sulfate or other compounds in place of oxygen.
Runaway greenhouse effect – an extreme water vapor feedback which results in boiling dry of the oceans. This happened on Venus, but is not expected to happen on Earth.
Scattering – reflection of light by cloud droplets. This differs from absorption in that the energy of the light is not absorbed, and the frequency of the light is unchanged.
Sea ice – ice that forms from freezing seawater at the surface of the ocean.
Silicate rock – an igneous rock. These consume CO_2 when they are weathered.
Silicate weathering thermostat – a negative feedback that stabilizes atmospheric CO_2 concentration on timescales of half a million years.
Skin altitude – the average altitude from which IR energy escapes to space.
Skin temperature – the temperature, on average, at which the Earth radiates IR light to space.
Soil carbon pool – dead organic carbon, mostly humic acid material. The soil carbon pool contains about 1500 Gton of C.
Storm surge – the low pressure in a hurricane or other storm lifts up the sea surface, like a very high tide.
Stratified – a fluid column with denser fluid on the bottom and less-dense fluid on top.
Stratosphere – a region of atmosphere above the troposphere, from 10 to about 50 km. The stratosphere contains about 10% of the air in the atmosphere.
Stratus clouds – layered clouds, formed by broad, gradual upward motion of air.
Sulfate aerosols – when sulfur is released to the atmosphere from combustion, say of high-sulfur coal, it forms tiny particles of sulfuric acid. These scatter light and act as cloud condensation nuclei.
Terawatt – 10^{12} W, abbreviated as TW. Humankind is currently producing and using 13 TW of energy.
Terrestrial biosphere – living carbon on land, mostly (by weight) plants. The terrestrial biosphere contains about 500 Gton of C.
Tragedy of the commons – a situation arising when the benefits of a common resource accrue to the individual, but the costs of its degradation are shared by all.
Transient climate response – a standard benchmark for comparing the time-dependent response of climate models to increasing CO_2. The TCR is the temperature when doubled CO_2 is reached, after a ramp-up of 1% per year.
Tree ring records – trees grow faster when the temperature is warmer, so the thickness of tree rings is used as a proxy for past temperature variations.
Tropopause – the boundary between the troposphere and the stratosphere.
Troposphere – the lower part of the atmosphere, from the ground to about 10 km. The troposphere contains 90% of the air in the atmosphere and is where weather happens.
Turbulent cascade – fluid in the atmosphere and ocean flows in eddies or gusts that range in size all the way from the global patterns down to sizes of centimeters or millimeters. Turbulent cascade is the flow of energy from large scales to small, as large eddies break up into smaller ones.

Ultraviolet light – abbreviated UV. Higher frequency light than visible, capable of causing sunburn or worse biological damage.
Urban heat island effect – paved cities are warmer than surrounding countryside.
Viscosity – the resistance against flow caused by friction in a fluid. Molasses is a high-viscosity fluid with lots of internal friction.
Volcanic CO_2 degassing – CO_2, either from metamorphic decarbonation or juvenile, that escapes from the Earth to the atmosphere.
Water vapor feedback – an initial warming allows more water to evaporate into the atmosphere. Water vapor is a greenhouse gas, so increasing its concentration leads to further warming.
Watt – a unit of energy flow, defined as joules per second.
Weathering – a chemical reaction wherein a rock, or some of the atoms in a rock, dissolve in water.

Constants and Symbols

c – speed of light in a vacuum, equal to 3×10^8 m/s.
e – a number used in the exponential function, e^x. The value of e is approximately 2.718.
n – wave number, defined as the number of waves per centimeter, often used to describe IR light.
α – Greek letter alpha used here to denote albedo.
ΔT_{2x} – Δ is the Greek letter delta, often used to denote a difference. In this case, it is the difference between two temperatures, an initial and a final after doubling CO_2.
λ – the Greek letter lambda denoting wavelength, the distance between successive peaks of a wave.
ν – the Greek letter nu, which denotes frequency.
σ – the Greek letter sigma, used to denote the Stefan–Boltzmann constant in the Stefan–Boltzmann equation (Chapter 2).

Index